THE TRISECTORS

UNDERWOOD DUDLEY

ZZ(TOP)

THE TRISECTORS

UNDERWOOD DUDLEY

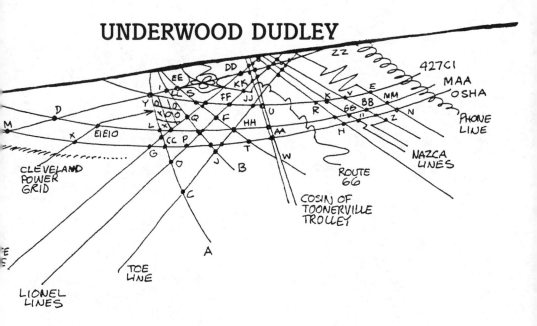

 The Mathematical Association of America

Revised edition of *The Budget of Trisections,*
©1987 Springer-Verlag.

©1994 by
The Mathematical Association of America (Incorporated)

Library of Congress Catalog Card Number 94-75575
ISBN 0-88385-514-3

Printed in the United States of America

Current Printing (last digit):
10 9 8 7 6 5 4 3 2 1

INTRODUCTION

My opinion of mankind is founded upon the mournful fact that, so
far as I can see, they find within themselves the means of believing in
a thousand times as much as there is to believe in.
          (Augustus De Morgan, *A Budget of Paradoxes,* vol. 1, p. 115)

In 1955, the mathematics library at the Carnegie Institute of Technology
was in a room at the end of a long, long hall. For much of its length, the hall
sloped slightly downward, either to follow the slope of the land underneath or
to show what engineers and architects could do if they set their minds to it. The
library was in a former classroom; it had three rows of shelves and four tables.
Borrowers of books were on their honor to write on a card the name of a book
taken out and the date it was taken. They were also on their honor to return it
within two weeks. Now the Carnegie Institute of Technology is the Carnegie-
Mellon University, and its mathematics library no doubt occupies much more
space and has many more books. I suspect that the honor system is no longer
in use.

Then, almost 40 years ago, the library was a pleasant place to be in the late
afternoon, with classes over for the day and the setting sun coming in the two
west-facing windows, lighting up the particles of dust in the air. Then as now,
students of mathematics (I was a student of mathematics then) generally did
not read anything that they were not required to read, so I was usually alone
there, with no one to tell me to take my feet off the table or to ask whether I
had some homework to do. I liked, and still do, to have my feet on a table, I
was talented enough then not to have to struggle with my homework as much

as most of my fellow students of mathematics, and, then as now, books were good company.

The library had a small and miscellaneous collection. Looking back, I see that it may have been intended as a student library, with the research library for the faculty located elsewhere, since there were hardly any of the rows and rows of bound periodicals so necessary for the researcher. On the other hand, there may have been no research library, for why would a student library have more than forty volumes of the Collected Works of Euler, mostly paperbound and mostly in Latin? In any event, on the bottom shelf on the right of the first row of shelves were two volumes, bound in red and black, that one day caught my eye: *A Budget of Paradoxes* they were called, written by Augustus De Morgan. Paradoxes—I thought that those were "proofs" that $2 = 1$, achieved by dividing by zero; or that every triangle is isosceles, done by drawing a diagram incorrectly; or that $i = -1$, accomplished by forgetting about a property of square roots: how could there be two large volumes devoted to them? I picked up the first volume and found that what was in it was not like that at all. It had some very strange things in it, written about in a style that I had never before encountered. The first edition was published in London in 1872, and the contents were contributions, to various periodicals, together with additions, that De Morgan had made over the years before his death in 1871.

De Morgan was a mathematician, a teacher of mathematics, a prolific writer, and an original man. Here is a portion of his nonmathematical biography, by J. M. Dubbey, in the *Dictionary of Scientific Biography*:

DE MORGAN, AUGUSTUS (b. Madura, Madras presidency, India, June 1806; d. London, England, 18 March 1871), *mathematics*.

De Morgan's father was a colonel in the Indian Army; and his mother was a friend of Abraham de Moivre, and granddaughter of James Dodson, author of the *Mathematical Canon*. At the age of seven months De Morgan was brought to England, where his family settled first at Worcester and then at Taunton. He attended a succession of private schools at which he acquired a mastery of Latin, Greek, and Hebrew and a strong interest in mathematics before the age of fourteen. He also acquired an intense dislike for cramming, examinations, and orthodox theology.

De Morgan entered Trinity College, Cambridge, in February 1823 and placed first in the first-class division in his senior year; he was disappointed, however, to graduate only as fourth wrangler in 1827. After contemplating a career in either medicine or law, De Morgan successfully applied for the chair of mathematics at the newly formed University College, London, in 1828 on the strong recommendation of his former tutors, who included Airy and Peacock. When, in 1831, the college council dismissed the professor of anatomy without giving reasons, he immediately resigned on principle. He resumed in 1836, on the accidental death of his successor, and remained there until a second resignation in 1866.

De Morgan's life was characterized by powerful religious convictions. While admitting a personal faith in Jesus Christ, he abhorred any suspicion of hypocrisy or sectarianism and on these grounds refused an M. A., a fellowship at Cambridge, and ordination. In 1837 he married Sophia Elizabeth Frend, who wrote his biography in 1882. De Morgan was never wealthy; and his researches into all branches of knowledge, together with his prolific output of writing, left little time for social or family life. However, he was well known for his humor, range of knowledge, and sweetness of disposition.

De Morgan wrote influential texts, his pupils included the well-known mathematicians Sylvester and Todhunter, he invented the term *mathematical induction,* he did fundamental work in logic, and he wrote *Arithmetical Books,* which was "probably the first significant work of scientific bibliography."

De Morgan's peripheral mathematical interests included . . . a curious work entitled *A Budget of Paradoxes,* which considers, among other things, the work of would-be circle squarers.

The library's copy of *A Budget of Paradoxes* was an edition published in Chicago, in 1915. It had very extensive notes added by David Eugene Smith, a historian of mathematics, because, he said,

> Hundreds of names are referred to in the text that were more or less known in England half a century ago, but are now forgotten and were never familiar elsewhere. Many books that were then current have now passed out of memory, and much that agitated England in De Morgan's time seems now like ancient history.

Smith provided a short paragraph for *every* name mentioned, and sometimes the notes filled half the page. They constituted an impressive feat of scholarship and were often as fascinating as the text.

The text was a curious mixture. Most of it I could understand, but there were parts which I could not penetrate on first or second reading. With repeated efforts, I have now mastered them all. De Morgan wrote with unfailing clarity, but a separation of five thousand miles of space and one hundred years of time can make things hard to see. *A Budget of Paradoxes* is full of wit, and it is full of high seriousness too; De Morgan is sometimes exasperated, sometimes genial, sometimes very earnest. By paradox, De Morgan meant anything counter to general opinion, and he defined a paradoxer as one who puts forth paradox. Today, acupuncture (for one example of many possible) is a paradox, and those who write booklets with diagrams of the body indicating where the needles should be stuck are paradoxers. Paradox is something not orthodox, so paradoxers are not necessarily wrong in what they maintain. Not necessarily, but usually they are. The *Budget* deals mostly with paradoxes in mathematics, physics, and religion, but it also contains digressions, sometimes quite long, on topics of amazing variety. There are anagrams of *Augustus De Morgan*—"Great gun Do us a sum!" is one (*Budget,* vol. 1, p. 138)—an Astronomer's Drinking Song:

> Copernicus, that learned wight
>   the glory of his nation,
> with draughts of wine refreshed his sight,
>   and saw the earth's rotation;
> Each planet then its orb described
>   the moon got under way, sir;
> these truths from nature he imbibed
>   for he drank his bottle a day, sir!

(Budget, vol. 1, p. 380)

and so on—many book reviews, such as

The Decimal System as a Whole. By Dover Statter. London and Liverpool, 1856.

The proposition is to make everything decimal. The day, now 24 hours, is to be made 10 hours. The year is to have ten months, Unsuber, Duober, etc. Fortunately there are ten commandments, so there will be neither addition to, nor deduction from, the moral law. But the twelve apostles! Even rejecting Judas, there is a whole apostle of difficulty. These points the author does not touch.

(Budget, vol. 2, p. 80)

And on and on. I read on and on, even through the parts I could not comprehend. In 1954, Dover Publications had reprinted the 1915 edition and a department store in Pittsburgh stocked a copy. Even though it was terrifically expensive ($4.95) I bought it and I am glad that I did.

Sometimes a book can give you the illusion that you know its author, that you know what he was like and how he behaved. Some biographies can do the same thing—I know Gauss—although other personalities are forever beyond reach—no one will *ever* know Newton. It is easy to become acquainted with Augustus De Morgan. It is no surprise to read in his wife's biography of him

After we were settled at No. 41 Chalcot Villas, Adelaide Road (at that time nearly surrounded by fields, and fully two miles from the College), he left the house always before eight o'clock in the morning, and met the omnibus in the Hampstead Road, which took him to Grafton Street a short time before the lecture began. He returned to dinner at five o'clock; and as he only gave himself about half an hour's rest after dinner before going to his library, where he wrote or read for four or five hours, he seldom gave up an evening to friends without feeling that his work for the next day had accumulated.

Nor is it a surprise to read a letter from De Morgan to a friend, in 1869,

You think, one letter of yours says, that I am feeling the effects of hard work; in fact, that I have been working too hard. Rid your mind of the idea. I have never been hard working, but I have been very continuously at work. I have never sought relaxation. And why? Because it would have killed me. Amusement is real hard work to me. To relax is to forage about the books with no particular object, and not bound to go with anything.

Yes, that is exactly the author of *A Budget of Paradoxes*.

A paradoxer could be anyone from Galileo, with his fearfully unorthodox ideas about the motion of the earth, to someone with a trisection of the angle with straightedge and compass alone. Some things that were once paradox are now orthodox: when Wegener advanced his idea of continental drift early in this century he got not acceptance but laughs and sneers. Today, if anyone doubts the truth of plate tectonics the doubter gets the laughs and sneers. There is, however, no chance whatever that status of the angle trisector will ever change. Such people, paradoxers who are demonstrably wrong, I will call *cranks*. I will apply the same term to people who maintain positions that, though not demonstrably wrong, have in almost everyone's opinion a very, very low probability of being correct. It would be impossible to *prove* that the Pyramidologists—that group, not as large as it once was, that asserts that the builders of the Great Pyramid of Cheops incorporated into it information about the future of the race, discoverable by taking measurements of it—are wrong, but cranks is what they are. Believers in the sunken continent of Atlantis, in the prophesies of Nostradamus, or in the existence of unidentified flying objects piloted by small green aliens, members of the Flat Earth Society (if there are any left), channelers and those who patronize them—cranks all, and all deserving of the title. *A Budget of Paradoxes* contains quite a few cranks. There were many circle-squarers, people who thought that they could construct a square with exactly the same area as a circle using straightedge and compass alone. And a pencil, of course. This amounts to the same thing as determining the value of π, the ratio of the circumference of a circle to its diameter. Many of De Morgan's circle-squarers claimed that the value of π was a rational number, such as the popular 3 1/8. It had been known since 1761, to mathematicians though not to paradoxers, that it was impossible for π to be a rational number. It had been proved to be impossible, and that was the end of it. Those circle-squarers were cranks. Other circle-squarers gave constructions which led to irrational values of π, but it was not until 1877 that it was finally proved that no straightedge and compass construction could ever succeed in squaring the circle. Until then, there was a hope that a construction could be found, but those circle-squarers with constructions which led to wrong values of π were cranks too.

Most of the mathematical cranks in *A Budget of Paradoxes* are circle-squarers. Circle-squaring was very popular in the nineteenth century. Circles were squared in books, pamphlets, single sheets, and even newspapers. Nowadays very few people square circles. There are trisections in newspapers now and then, but no circle-squarings. I like to think that the *Budget* was the main cause in the decline. As De Morgan says,

> If I had before me a fly and an elephant, having never seen more than one such magnitude of either kind; and if the fly were to endeavor to persuade me that he was larger than the elephant, I might possibly be placed in a difficulty. The apparently little creature might use such arguments about the effect of distance, and might appeal to such laws of sight and hearing as I, if unlearned in those things, might be wholly unable to reject. But if there were a thousand flies, all buzzing, to appearance, about the great creature; and, to a fly, declaring, each one for himself, that he was bigger than the quadruped; and all giving different and frequently contradictory reasons; and each one despising and opposing the reasons of the others I should feel quite at my ease. I should certainly say, My little friends, the case of each of you is destroyed by the rest. I intend to show flies in the swarm, with a few larger animals, for reasons to be given.   (*Budget,* vol. 1, p. 1)

There once was a swarm of circle-squarers, but there is one no longer. Perhaps De Morgan caused that, and I hope that this book will have a similar effect on angle trisectors.

For whatever reason, De Morgan and cranks gripped me and I have been gripped ever since. Whenever I came across a piece of crank literature, I held onto it. I began to try to search cranks out. I put an advertisement in *Fate* magazine, at the time a sort of *Reader's Digest* of the occult. I went to the Library of Congress and copied all that I could find. I wrote to six hundred departments of mathematics, asking whether anyone there were a fellow collector or, if not, whether they had any material on file. No one would admit to collecting it. It was hard to believe: in a country where there are collectors of barbed wire and of telephone insulators, is there no one collecting crank mathematics? Evidently not, though many departments were generous in sending material. But many other departments had none at all. Some departments throw such stuff away immediately. Other departments file all of their correspondence, and that from cranks goes into a folder labeled *Nuts, Crackpots,* or something similar, though that label is not fair. What De Morgan wrote is still true:

> They are in all ranks and occupations, of all ages and characters. They are very earnest people, and their purpose is *bona fide* the dissemination of their paradoxes. A great many—the mass, indeed—are illiterate, and a great many waste their means and are in or approaching penury. (*Budget,* vol. 1, p. 8)

It may be that the illiteracy rate is lower now. Or perhaps De Morgan was counting the writer of the following letter (I have received similar ones) illiterate:

> When a Gentleman of your standing in Society Clad with those honors Can not understand or Solve a problem That is explicitly explained by words and Letters and mathematically operated by figures He had best consult the wise proverd
>
> Do that which thou Canst understand and Comprehend for thy good. I would recommend that Such Gentleman Change his business
>
> And appropriate his time and attention to a Sunday School to Learn what he Could and keep the Little Children from durting their Close.
>
> With Sincere feelings of Gratitude for your weakness and Inability I am, Sir your superior in Mathematics
>
> (*Budget,* vol. 2, pp. 16–17)

The trouble is that every department sooner or later purges its files and the crackpot folder is always destroyed. So this material, the folk mathematics of the time, is lost forever. It is a shame, for crank mathematics is worth at least as much attention as many things to which scholars pay attention. If, reader, you know of any mathematical crank literature, I would be pleased to have it, or a copy of it: I have the collector's lust.

Crank mathematics is mostly produced by amateurs, doing it for fun, or for the challenge, or for the fictitious million dollars they have heard is the reward for solving some problem. There are a few cases, some notorious in the mathematical community, of professional mathematicians turning into cranks (or displaying their crankhood, if you believe as I do that cranks are born and not made), but not many. You might think that all angle trisecting is done by amateurs since professionals understand what "impossible" means in mathematics, but there is an exception; in the Budget of Trisections which is the last section of this book, there appears a trisector with a Ph. D. degree in mathematics who earned his living teaching mathematics. It is incredible that such a person should be a trisector, but there he is. But almost all crank mathematics is done by amateurs: somewhere, sometime, a person remembers that time in tenth-grade geometry when the teacher said that it is impossible to trisect angles with straightedge and compass alone. Then, not knowing what "impossible" means in mathematics, he gets out compass and straightedge and starts to attack the problem. He is on his way. It is fun. He finds a construction that looks very good for an angle of $30°$, giving $10°$ so closely that not even a big protractor can tell the difference, but which doesn't look quite right for a $75°$ angle. How to modify the construction so that the trisection point will be a little more in that direction? Maybe if I draw *this* circle instead of *that* one. It is fun: problems presented, problems solved, small triumphs and small failures, all together with the sensual pleasure of drawing neat diagrams with sharpened pencils. The feel of a nicely balanced sharp-pointed compass! The crisp intersection of line and line! The tangent line, grazing the circle at precisely one point! What better way to spend an evening?

There are many better ways. The fun stops when the construction is completed and then the frustration begins. Mathematicians will not look at the construction. Or they look at it and use trigonometry, of which the average trisector is ignorant, to show the trisector that he is wrong. Or they give him proofs, which he cannot understand, that the trisection is impossible. All of them, over and over, say he is wrong, wrong, wrong. What was pleasure turns to pain. It is too bad. The purpose of this book is to reduce the amount of that pain. If a trisector, or a potential trisector, can see the swarm in the Budget of Trisections, perhaps he will turn from the trisection to some other form of recreational mathematics. That will be pure gain, both for him and for the mathematical community. I have the opinion, based on no evidence whatsoever, that there is a crank personality and that some people are destined to be cranks, so perhaps he will become a physics crank, refuting Einstein; no gain for him, but gain for the mathematical community, which will no longer be bothered with him. Of course, if he starts to square the circle, no one has gained, but we can only try our best.

By the way, the trisection *is* impossible. It was proved so by Wantzel, in 1837. That is all the typical history of mathematics has to say about Wantzel, if it mentions him at all. He deserves more, especially since he also gave the first proof of a result about regular polygons that the great Gauss said that he had proved but that he had never published. Pierre Laurent Wantzel was, as his first names show, French; he was born in 1814; he was a talented mathematician; he published his proof in Liouville's *Journal de Math.* (vol. 2, 1837, pp. 366–372); and he died in his thirty-fifth year. A contemporary wrote

> Ordinarily he worked evenings, not lying down until late; he then read, and took only a few hours of troubled sleep, making alternately wrong use of coffee and opium, and taking his meals at irregular hours. He put unlimited trust in his constitution, very strong by nature, which he taunted at pleasure by all sorts of abuse. He brought sadness to those who mourn his premature death.

The idea of Wantzel's proof is that constructions with straightedge and compass allow one to perform with lines and circles the arithmetical operations of addition, subtraction, multiplication, division, taking a square root, and no others, whereas to trisect an angle it is necessary, in effect, to take a cube root. There is no way of combining any number of plusses, minuses, timeses, divides, and square roots to get a cube root, and that is why the trisection cannot be done with straightedge and compass alone. The proof uses nothing more advanced than algebra, but it is just hard enough and just involved enough to make it impossible for anyone without a good deal of mathematical training to understand it. Even so, well-meaning but not clear-thinking professors of mathematics continue to send copies of the proof to trisectors. I have not included the proof here because if you have the background necessary to understand it, you know already that the trisection is impossible and you do not need to read it, and if you lack the background, it would be too hard for you to understand. Besides, it is easy to find so there is no need to reproduce it here. W. V. Quine, the eminent logician and philosopher, published "Elementary proof that some angles cannot be trisected by ruler and compass" in *Mathematics Magazine* 63 (1990) #2, 95–105, and his clear and elegant prose can hardly be improved on.

Hardly any mathematical training is necessary to read this book. There is a little trigonometry here and there, but it may be safely skipped. There are only a few equations. There are no exercises at the end of the sections, and there will be no final examination. The worst victim of mathematics anxiety can read this book with profit and dry palms. It is suitable for giving as a present.

There is hardly any literature on the trisection. Every now and then a journal will print an approximate trisection, and the proof that it is impossible appears here and there, but the only book I know devoted to it is *The Trisection Problem,* by R. C. Yates, a slim volume, published by the National Council of Teachers of Mathematics, mostly concerned with the proof of impossibility and with constructions using compass, straightedge, and something else to accomplish the trisection. Trisectors are mentioned hardly at all.

What follows, then, is something which has never been done before: it is an effort to do something which may be as impossible as trisecting the angle: namely to put an end to trisections and trisectors. It was inspired by *A Budget of Paradoxes* but lacks its wide scope, and its author attempts the manner of De Morgan but lacks his vast erudition. There were giants in those days.

Addendum for the second edition

Well, it didn't work. The publication of the first edition of this book did not put a stop to trisectors and trisections. It did not even make a dent. Trisections continue to appear. There was even a United States Patent issued on May 18, 1993 (number 5,210,951) for a trisecting device, the first trisection patent issued in this country since number 3,906,638 in 1973. What is worse, I have received more than one letter whose writers said, in more or less these words

> I have read your book on trisections. It is very interesting. Here is my method for trisecting angles using only straightedge and compasses.

Perhaps I was not sufficiently clear. Let me try one more time:

YOU CAN'T TRISECT ANGLES!
DON'T TRY!

There. Maybe that will do it.

I have not added any new constructions to this edition since the new ones are much like the old. There is a small amount of new material and the old material has been revised in small ways. Many of the errors in the first edition have been corrected and, unfortunately but inevitably, brand-new errors have been added to this edition, though not on purpose. The error-free book, like the trisection-free world, is an unattainable ideal.

Contents

Chapter 1

Non-Euclidean Constructions

Always write in your books. You may be a silly person—for though your reading my book is rather a contrary presumption, yet it is not conclusive—and your observations may be silly or irrelevant, but you cannot tell what use they may be of long after you are gone where Budgeteers cease from troubling.   (*Budget,* vol. 2, p. 265)

Where would we be without the Greeks? The ancient Greeks, that is, the only people in the ancient world who had the idea of using reason to solve problems and reach decisions. Egypt, Babylonia, China—high civilizations, but things were not done for reasons there; they were done because they had always been done, or because the priest said to do them, or because the king so ordered. The Rhind papyrus, that ancient Egyptian textbook of mathematics, is nothing but a series of orders: to solve this problem, do this, and this, and then this. The author may have known why the rules worked and not written down the reasons because it was not the style of the time, but it is also possible that the rules were discovered by accident, the finder stumbling on something which always gave the right answer for unknown reasons. Ancient Babylonian mathematical clay tablets are much the same.

But not the Greeks! They *reasoned.* They had debates. They got at truth by using logic. Not for nothing has this been called the "Greek Miracle." If the Greeks had not invented reason, would anyone else have? Not the Babylonians, and certainly not the Egyptians. The Romans might have, but then again they might not, and then where would the world be? Probably in much the same state as 2,500 years ago.

It was an exciting time to be alive, at the dawn of rationality. Though they did not find the answers, the ancient Greeks asked all of the questions. What is the nature of the universe? What is the purpose of life? How shall society be organized? How shall citizens conduct themselves? Why is the square of the hypotenuse of a right triangle the sum of the squares of the other two sides? For the first time, reason was being used to try to get at the answers. Reason was still mixed with older modes of thought, and when we read that the universe was all water or, as the Pythagoreans said, all number, we do not feel ourselves in the presence of keenly reasoning minds.

It was in geometry that reason had its greatest early success, so great that Euclid's *Elements of Geometry,* which summarized geometrical knowledge around 300 B.C., was used as a textbook for almost 2,000 years. Geometry showed the immense power of reason. Starting with only five postulates, which in effect said

1. A straight line can be drawn between two points.
2. Lines can be indefinitely extended.
3. A circle can be drawn with any center and radius.
4. All right angles are equal.
5. Through a point outside a line, one and only one parallel can be drawn to the line.

and a few common notions—if two things are both equal to a third thing, they are equal to another, for example—it was possible to deduce many, many theorems. The sum of the angles of a triangle is two right angles. The square of the hypotenuse really is the sum of the squares of the other two sides, not because it has always worked out that way for all of the triangles anyone has ever looked at, but because it can be *proved*: deduced from statements that everyone agreed were true. From truth, only truth can be deduced. And, better yet, geometry gave truths about the world. Physical right triangles behaved in the way the Pythagorean theorem said they would. Fields in the shape of trapezoids had the areas that Euclid said they would. A triumph for reason! It was too bad that in the nineteenth century it became clear that Euclidean geometry does not in fact have any necessary relation with physical reality, but that is another matter entirely.

The ancient Greek geometers, who, even though they lived long ago were every bit as smart as we are (and then some), were probably convinced that it was not possible to trisect an angle with straightedge and compass, though the proof of that would have to wait for many centuries. They no doubt concluded that if there were a way that the trisection could be done, someone would have found it already. After all, problems that sound much harder had been solved, such as finding the area enclosed by a parabola and a line or finding a circle tangent to three given circles.

The three famous problems which could not solved with straightedge and compass alone—the trisection of the angle, the squaring of the circle (finding a square with the same area as a circle), and the duplication of the cube (finding a cube with double the volume of a given cube)—seem to have been something of an obsession with the Greeks. Each generation of geometers attacked them anew and solved them in new ways. I will present a few of the ways of trisecting angles, some from the ancients and some more modern, to show the wide variety of methods.

This material is not new, and much of it is readily available elsewhere, so mathematicians of wide learning are entitled to sneer at my copying what has been done before. The reason for including it is that many trisectors do not know that exact trisections are easy to do. They are under the delusion that there is a *need* for a method for trisecting. They are unaware that dividing an angle into three equal parts may be quickly accomplished with a protractor, available at any up-to-date drugstore. Nor does the art and science of geometry feel the lack. Look into *Mathematical Reviews* to see what geometers are worrying about and you find papers with titles like

> A note on finite subfield planes that admit affine homologies
> The nonexistence of an oval-extendable (56, 11, 2) design
> On a generalization of Kantor's likeable planes
> On the type of partial t-spreads in finite projective spaces
> Divisible semibiplanes and conics of Desarguesian biaffine planes

all taken from a recent issue. Do not ask me what a likeable plane is, because I do not know. I doubt that semibiplanes can fly. But the thing to note is that there is not an angle in sight. Geometry has passed them by.

Another reason for showing these exact trisections is that they may persuade a trisector or two that his construction is not exact: if they find that their constructions do not agree with an accurate trisection, they may stop trisecting. At least a seed of doubt may be planted. Besides, the pictures are pretty.

First, because he was first among mathematicians, is the trisection by the great Archimedes (287–217 B.C.). Engineer, physicist, mathematician, one of the greatest minds humanity has produced, Archimedes no doubt threw off this trisection as a trifle, the work of an evening. It is elegant and simple and uses the very minimum over the Euclidean tools of straightedge and compass. Archimedes trisected the angle, and you can too, using a compass and a straightedge with two scratches on it.

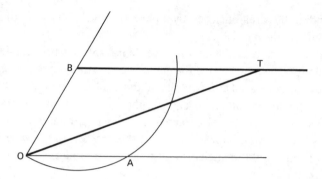

FIGURE 1.1

In Figure 1.1, take angle BOA to be trisected where $|BO|$ is the distance between the scratches on the straightedge. Draw a circle, center B, radius BO. Draw a line through B parallel to OA. Now put the straightedge through O so that one of the two scratches is on the circle at C and the other is on the line through B, at T. T is then the exact trisection point T. Hardly anything to it! Easy to do if you have a parallel rule, and not much harder if you don't: to draw the parallel, drop a perpendicular from B to OA and then erect a perpendicular to that line at B.

To see why the trisection is accurate is not hard, either. In Figure 1.2, let θ be the measure of angle COA. Because BT is parallel to OA, angle CTB is also θ ("When a transversal cuts a pair of parallel lines, the alternate interior angles are equal.") Triangle BTC is isosceles, because $|BC| = |BO|$ ("All radii of a circle are equal.") and $|CT| = |BO|$ because of the way we placed the marked straightedge. Since the triangle is isosceles, angle CBT is also θ. ("The base angles of an isosceles triangle are equal.") This makes angle BCT equal to

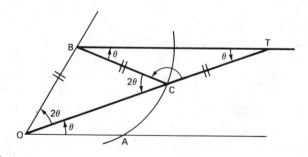

FIGURE 1.2

$180° - 2\pi$ ("The sum of the angles of a triangle is two right angles.") So angle OCB is 2θ ("The sum of two supplementary angles is $180°$.") Triangle OBC is isosceles too, because $|OB| = |BC|$, so angle BOC is also 2θ. This shows that angle BOA, 3θ to start with, has been trisected by OT.

This elegant construction has been rediscovered more than once, and more than one trisector has tried to locate point C with straightedge and compass. An improvement was published in 1907 (E. E. White, "On the trisection of an angle," *American Mathematical Monthly* 14 (1907), 151–152); for this construction a straightedge with two scratches $2|BO|$ apart is needed. In Figure 1.3, drop a perpendicular from B to A and draw a line through B parallel to OA. Then place the straightedge so that it goes through O, one scratch on BA and the other on the line through B. That will be the trisection point.

The relation to the Archimedean trisection is shown in Figure 1.4, where I have drawn in the construction lines that would be needed to give a proof. With one hint ("The diagonals of a rectangle have equal length and bisect each other."), I will leave to the reader the task of proving that the trisection is exact.

It is possible to make a mechanical device to do the Archimedean trisection. It consists of four rods, jointed at O and C, with a sliding joint at A and

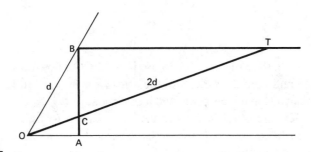

FIGURE **1.3**

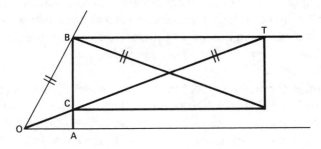

FIGURE **1.4**

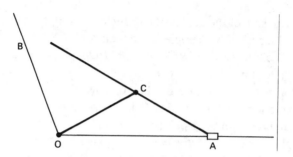

FIGURE 1.5

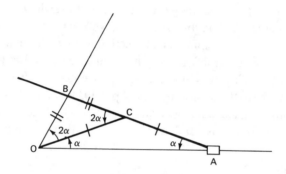

FIGURE 1.6

with $|OC| = |CA|$. (See Figure 1.5.) To use it, hold BOA down over the angle to be trisected and slide the slide until the arms cross at equal distances from C and O. Then you have Figure 1.6, and a trisection.

Archimedes flourished in the century after Euclid wrote his *Elements*. Long before Euclid there were trisections; the earliest known was done by Hippias, who was born sometime around 460 B.C. Hippias seems to have been one of those unpleasant mathematicians: Plato quoted him as boasting that he had made more money than any two other Sophists put together and Plato called him boastful of his knowledge. He probably had something to boast about, because he was clever enough to invent a curve which could be used to divide an angle into any number of equal pieces and which would square the circle as well.

To construct the Quadratrix of Hippias, in Figure 1.7 move segment DE down, parallel to BC, at a uniform rate and move radius OR clockwise at a uniform rate so that they both reach OA at the same time. In the figure, DE is one-third of the way down, OR is one-third of the way around, and Q is one point on the Quadratrix. The path of Q, the intersection of the segment and

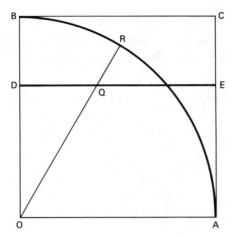

FIGURE 1.7

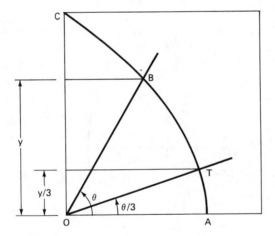

FIGURE 1.8

the radius, is the Quadratrix; it looks like the curve CA in Figure 1.8. Given the curve, the trisection is easy, since the distance down is proportional to the rotation around. So, in Figure 1.8, y and θ are proportional; to divide θ by 3, divide y by 3.

A plot of the curve can be gotten from a computer or by using a pocket calculator. If the square has side 1 and O is the origin of a rectangular coordinate system, then the coordinates of B, (x, y), satisfy $\tan \theta = y/x$. But θ and y are proportional: in fact, $\theta = \pi y/2$. So, $\tan(\pi y/2) = y/x$ or

$$x = y \cot(\pi y/2).$$

It is not possible to solve for y in terms of x in any convenient way, but by substituting in values for y from 1 down to 0, you can plot as accurate a quadratrix as you want.

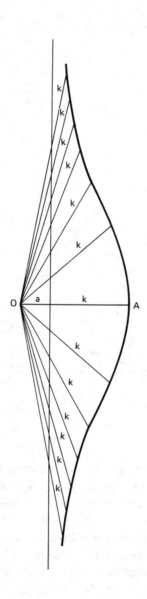

FIGURE 1.9

Let me include another Greek curve that will do the trisection, the Conchoid of Nicomedes. Does not the name of that curve have a ring to it? It is too bad that the good old curves—the Cissoid of Diocles, the Lemniscate of Bernoulli, Tschirnhausen's Cubic, the Witch of Agnesi, the Trochoid, the Astroid, the Deltoid, the many Roulettes, the Folium of Descartes—have fallen out of mathematical fashion and are neither much studied nor taught these days. To make a conchoid, take a line, a point, and a distance, k, and add the distance to all the segments from the point to the line. The ends of the augmented segments lie on the conchoid, a pleasing curve with a shape like that of the normal density in probability theory (see Figure 1.9).

To draw a conchoid by plotting points, use either the rectangular equation

$$y^2 = \frac{x^2(k+a-x)(k-a+x)}{(x-a)^2}$$

or the polar equation

$$r = k + a\sec\theta.$$

To use the curve to trisect, place one side of the angle along OA and put the vertex in a place so that $|BO|$ is half of k, as in Figure 1.10. A line through B parallel to OA will intersect the conchoid at the trisection point, T. Figure 1.11 is a picture of a conchoid-drawing instrument; the two dots represent pegs that slide in the slots.

Another of Archimedes' inventions, his spiral, can be used as a trisecting tool. There is a picture of one in Figure 1.12; the distance from O to B

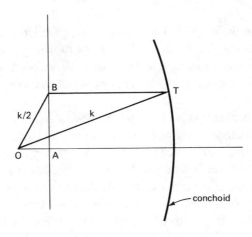

FIGURE 1.10

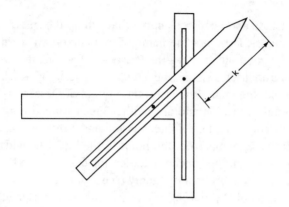

FIGURE 1.11

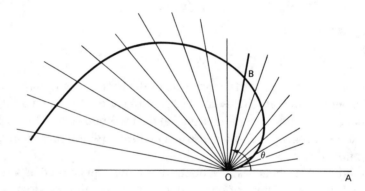

FIGURE 1.12

is proportional to the angle θ. The idea for the trisection is the same as for the Quadratrix of Hippias: because of the proportionality of the straight line distance and the angular measure, to trisect an angle, we need only trisect a segment. In Figure 1.13, to get one-third of angle BOA, get one-third of segment BO.

Although the trisection is no longer the obsession it was to Greek mathematicians, new methods have been devised in modern times. Etienne Pascal (1588–1640), the father of Blaise (1623–1662) the mathematician, philosopher, and writer, trisected with a cardioid. It is constructed (see Figure 1.14) with the same idea as the conchoid, except that instead of a point and a line we use a point and a semicircle and make k and a the same length. Once you have your cardioid with you, to trisect angle BOA (Figure 1.15), put it at the center of the circle and extend OB to C on the cardioid. Then angle OCD is

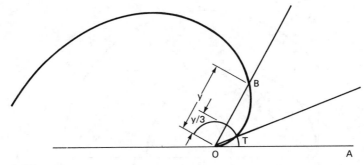

FIGURE 1.13

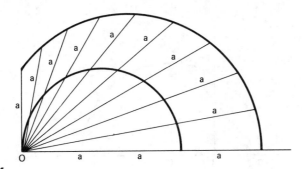

FIGURE 1.14

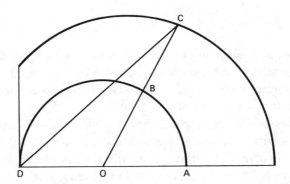

FIGURE 1.15

one third the original angle. Quickly done! To see why it works, in Figure 1.16
$|OD| = |OE| = |EC|$ because of the way the cardioid was made. Start with
angle EOC, θ degrees, and follow through the diagram to get, successively,
angle $OCE = \theta$ degrees, angle $OEC = 180° - 2\theta$, angle $OED = 2\theta$, angle
$ODE = 2\theta$, angle $DOE = 180° - 4\theta$, and finally angle $BOA = 3\theta$.

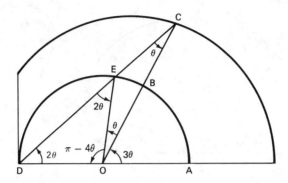

FIGURE 1.16

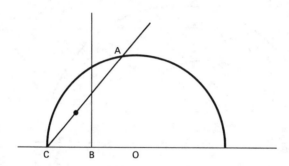

FIGURE 1.17

Nicomedes and Pascal: two minds with the same idea, first with a line and then, 2,000 years later, with a semicircle. Clearly, the idea could be used with any kind of curve: do it with a square and you will have devised a new trisecting curve, never before drawn (I think).

Let me include one more curve to show that it is not hard to get a trisection. This is the Trisectrix of Maclaurin (Colin Maclaurin, 1698–1746) and is easily drawn with a pair of dividers. Start with a semicircle and erect a perpendicular bisector of segment CO at B (Figure 1.17). For a given angle θ, the point on the trisectrix is along the ray at a distance $|CP|-|CQ|$ from O. In Figure 1.18 more points on the curve have been plotted. To trisect, put the angle to be trisected with its vertex at O, as in Figure 1.19, and connect B, the point on the trisectrix, with C and you are done; angle BCO is one-third angle BOA.

The mathematics of why the trisectrix works is considerably more complicated than for any of the previous curves. In polar coordinates, the equation of the semicircle is, if it has radius 1, $r = 2\cos\theta$ and the equation of the vertical line is $r\cos\theta = 1/2$. Thus for any θ, $|CP| = 2\cos\theta$, $|CQ| = 1/(2\cos\theta)$ and the

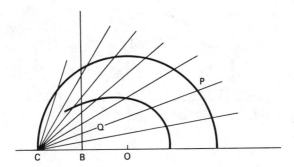

FIGURE 1.18

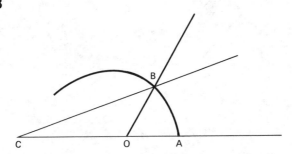

FIGURE 1.19

equation of the trisectrix is

$$r = |CP| - |CQ| = 2\cos\theta - \frac{1}{2\cos\theta}$$

which, with some algebraic manipulation, turns out to be equal to $(\sin 3\theta)/(\sin 2\theta)$. So, in Figure 20,

$$\tan\alpha = \frac{y}{x-1} = \frac{r\sin\theta}{r\cos\theta - 1}$$
$$= \frac{\sin 3\theta \sin\theta}{\sin 3\theta \cos\theta - \sin 2\theta}$$

which, with some more algebraic manipulation, turns out to be equal to $\tan 3\theta$. That is, $\alpha = 3\theta$ and θ is one-third α.

The equation of the trisectrix in rectangular coordinates is

$$y^2 = \frac{x^2(3-2x)}{1+2x}.$$

I will conclude with two mechanical trisectors, the carpenter's square and the shoemaker's knife. The carpenter's square trisection uses a square with one side two inches wide; the construction does not seem to go back any further

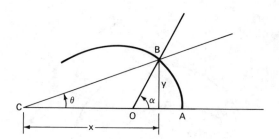

FIGURE 1.20

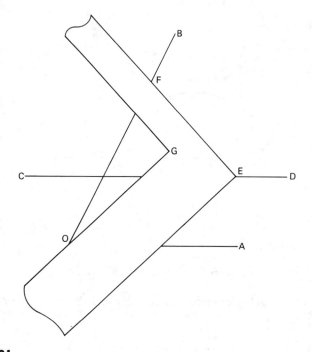

FIGURE 1.21

than 1928 (H. T. Scudder, "How to trisect an angle with a carpenter's square," *American Mathematical Monthly* 35 (1928), 250–251.) Given angle *BOA*, use the square to draw *CD* parallel to *OA* and two inches from it. Then place the square (Figure 1.21) so that *EF* is four inches long, *E* is on *CD*, and the inside edge of the square goes through *O*. Angle *FOG* is one-third of angle *BOA* because, in Figure 1.22, all three triangles are congruent.

The shoemaker's knife, or tomahawk, works in the same way. There is a picture of one in Figure 1.23; the curve is a portion of a semicircle. The de-

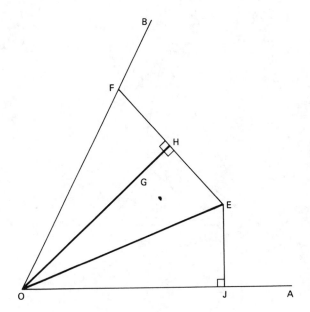

FIGURE 1.22

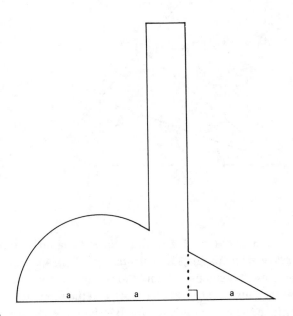

FIGURE 1.23

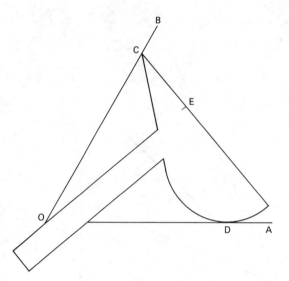

FIGURE 1.24

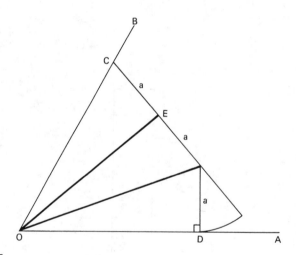

FIGURE 1.25

vice was invented in 1877 (M. H. Bricard, "Note sur la division mechanique de l'angle," *Bulletin de la Société Mathématique* 5 (1877), 43–47. Attributed to "M. Glatin."). To use it, maneuver it until the point is on one side of the angle, the vertex is on the outside edge (Figure 1.24), and the semicircle is tangent to the other side. Like the carpenter's square, the shoemaker's knife gives three congruent triangles (Figure 1.25), and so the trisection is accomplished.

Other novel ways to trisect can be found in the *Mathematics Teacher* (by A. E. Hochstein, 56 (1963), 522–524) using straightedge, compass, and semireflector, in *School Science and Mathematics* (by Floyd S. Lorentz, 47 (1947), 255–257) using straightedge and transparent paper alone, in *School Science and Mathematics* (14, (1914), 236), using a machine, and in the *Mathematics Teacher* (by Norman Anning, 44 (1951), 194–195), using compasses and T-square.

There have been many other mechanical trisectors. I will not lengthen the catalog, but will finish with an appropriate comment on all such things by Leo Moser (*Scripta Mathematica* 13 (1947), 57): to trisect an angle, all you need is a watch and patience. An old-fashioned watch with hands, that is; no one has yet shown how to accomplish the trisection with a digital watch. Set the watch at 12:00 and consider the direction of the hands as one side of the angle to be trisected, call it θ. When the minute hand reaches the other side of the angle, the hour hand is at $\theta/12$. Double this angle twice and you have $\theta/3$. See Figure 1.26.

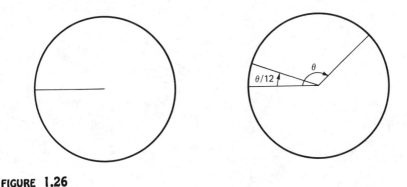

FIGURE 1.26

Now you can trisect an angle anytime, anyplace, for anyone who asks. But no one ever will.

Chapter 2

Characteristics of Trisectors

The feeling which tempts persons to this problem is that which, in romance, made it impossible for a knight to pass a castle which belonged to a giant or an enchanter. I once gave a lecture on the subject: a gentleman who was introduced to it by what I said remarked, loud enough to be heard by all around, "Only prove to me that it is impossible, and I will set about it this very evening."

This rinderpest of geometry cannot be cured, when once it has seated itself in the system: all that can be done is to apply what the learned call prophylactics to those who are yet sound. When once the virus gets into the brain, the victim goes round the flame like a moth; first one way and then the other, beginning where he ended, and ending where he began.   (*Budget,* vol. 2, p. 210.)

The trisector of an angle, if he demand attention from any mathematician, is bound to produce, from his construction, an expression for the sine or cosine of the third part of any angle, in terms of the sine or cosine of the angle itself, obtained by help of no higher than the square root. The mathematician knows that such a thing cannot be; but the trisector virtually says it can be, and is bound to produce it, to save time. This is the misfortune of most of the solvers of the celebrated problems, that they have not knowledge enough to present the consequences of their results by which they can be easily judged. Sometimes they have the knowledge and quibble out of the use of it. In many cases a person makes an honest beginning and presents what he is sure is a solution. By conference with others he at last feels uneasy, fears the light, and puts self-love in the way of it. Dishonesty sometimes follows. The speculators are, as a class, very apt to imagine that the mathematicians are in a fraudulent conspiracy against them: I ought to say that each one of them consents to the mode in which the others are treated, and fancies conspiracy against himself. The mania of conspiracy is a very curious subject.

(*Budget,* vol. 2, pp. 12–13.)

What is it that makes people become trisectors? That is a question which will never be answered, but it is possible to give some of the characteristics of trisectors. I will proceed to do this, so that you will be able to recognize one when you see him and not be surprised by what he does or says.

"Him" and "he" are correct: almost 100 percent of trisectors are male. There are only two females in the Budget of Trisections and, from what they wrote, it is possible that they are not genuine trisectors. They sounded a bit tentative and may not have known that the trisection is impossible. Men trisect; women do not. You can explain that by saying that women have too much sense to waste time on such foolishness as trisections, but that merely brings up the question of why women have sense and men do not. It is probably best not to pursue this question any further.

Trisectors are old. The typical trisector heard about the trisection in his high-school geometry class, years and years ago, but did not work on it until many years later, sometimes not until after retirement. Of course there are young trisectors too, who hear about the problem but fail to hear that it is impossible, but I think that they are all convinced when some authority tells them that they have not succeeded, unless, of course, they later turn out to be old trisectors. The trisection is like a disease with a very long incubation period. Trisectors tend to be old men. One wrote in 1953:

> I became interested in this problem in my junior high school year (1913–14) at Philo, Illinois. My geometry teacher suggested it. I turned it over in my mind and thought of it many times.

I wonder what the geometry teacher really said. Whatever it was, the teacher caused a new trisection to appear 40 years later. Here is an engineer writing in 1973:

> It all started in 1936. From that date on, more or less of my spare time was devoted in the field of trisection.

From a periodical in 1972:

> My teacher said that it was the opinion of mathematicians that a solution was impossible. For over 55 years that puzzle has bugged me.

"Opinion": did the teacher really say that? You would hope not, but what did the teacher say? A trisector in 1973:

In more than 12,000 working hours I have in the course of 40 years found this solution. I am not a mathematician but a retired civil servant, now 69 years of age.

Twelve thousand working hours! Full-time work is forty hours a week for fifty weeks of the year, or 2,000 hours a year. The poor man had spent the equivalent of six years of his life in trying to do the equivalent of finding two even integers whose sum was odd. What would you give for six extra years in your life? What would I? What couldn't we accomplish with that time? What a deplorable waste! Trisectors have always been old. There are only two trisections in *A Budget of Paradoxes,* and De Morgan's entire review of one of them is

"The consequence of years of intense thought"; very likely, and very sad. (*Budget,* vol. 2, p. 10.)

Another obvious characteristic of trisectors is that they fail to understand what "impossible" means in mathematics. The meaning is unfortunately not the same as the meaning in English, in which it can mean "too difficult to be accomplished now." Today it is a true statement that "It is impossible to land a person on any of Saturn's moons," but sometime in the future it may be false. When something is impossible in mathematics, it is impossible once and for all, and the passage of time will not change it. The sum of two even integers is even, yesterday, today, and tomorrow, forever and ever. It is one of the great failures of mathematics education that this essential difference is not made plain to students. (Only one of the great failures: the greatest failure is the failure of students to learn mathematics.) Part of the failure may be due to teachers' either not understanding the difference either or failing to mention it. In fact, both should share the blame. Typical is the trisector who wrote

I received through the mail an advertising brochure, from a science magazine, that had in it a simple statement and it went something like this—the FORMULA for TRISECTING AN ANGLE had never been worked out. This really intrigued me. I couldn't believe, after hundreds of years of math, that this could be true.

So he went to the library and found that all of the books agreed that it was impossible.

How could men of science be so stupid? Any scientist or mathematician who declares that a thing is impossible is *showing his limitations* before he even starts on the problem at hand.

A clear lack of understanding. Another trisector wrote in 1933:

> Moreover, we find our modern authorities of mathematics not at-
> tempting to solve these unsolved problems, but writing treatises show-
> ing the impossibility of proving them. Instead of offering inducements
> to the solution of these problems, they discourage others and dub
> them as "cranks."

This trisector had a testimonial from a professor of mathematics who I will not
name, though he is certainly dead by now, who taught at a real university which
I will also not name, since it is still alive:

> I have carefully checked your work on "the trisection of the angle"
> and was not able to detect any fallacy in your work.

There are two possibilities. Either the professor was genuinely unable to find
any flaw, which does not speak well for his mathematical capacity, or he hit on
this course as the quickest way to be rid of a pest, which does not speak well
for his ethical capacity. In either case it is not to his credit, nor to that of his
university. It is a good general rule never to give a testimonial to a trisector. It
is an even better one never to encourage a trisector, yet there are mathematicians who will do things which have exactly that effect. Here is a professor of
mathematics writing to a trisector:

> I looked at your diagrams briefly. My suggestion is that you concen-
> trate on just one angle 80°, 40°, 20° or 10° and write down exactly
> and in order the steps you take to construct the angle. Someone in
> your area might be willing to go over it with you. Good luck.

That is irresponsible. "Good luck" indeed! Of course, the aim was to get rid
of the trisector as quickly as possible with no hard feelings, but to do this by
encouraging folly is not right. Every effort should be made to get trisectors to
stop trisecting as soon as possible, since the longer they continue and the more
they invest in their constructions, the greater will be their eventual bitterness
and frustration. It was an angry trisector who wrote

> As you probably know there is little reward for the author of any sci-
> entific findings, the way the Copyright Laws are today. If I can't find
> a University to review the work of 1/2 a life time and if the Copy-
> right laws won't copyright until published and the publisher won't act
> without some recognized mathematician's approval, then I'm going
> to turn the matter over to Congress, and I might do it right away.

Anyone who would write, as did another mathematician,

I regret that I never quite succeeded in explaining the difference between an unsolved problem and a proof that, given certain tools, some constructions cannot be performed with those tools.

I wish you much success on your further researches and thank you for informing me of your findings.

should have his membership in the American Mathematical Society suspended, for two years at least. In some situations, there is no place for politeness.

A fourth characteristic of trisectors is that they do not know much mathematics. High school geometry is as far as most of them have gone, and some have not even gone that far. One trisector wrote in 1902

It was necessary to get outside of the problem to solve it, and it was not solved by a study of geometry and trigonometry, as the author has never made a study of these branches of learning.

What colossal presumption to think that he, ignorant of the subject, could solve a problem that as far as he knew had baffled the best efforts of the best minds over the centuries! What gall and nerve! Would he today try nuclear fusion in his basement? And think that he had succeeded? On the other hand, perhaps it is a tribute to the spirit of American democracy. Every person is exactly as good as every other in spite of heredity or environment.

You might think that anyone who knows higher mathematics could not be a trisector, and it is true that a knowledge of trigonometry seems to give some immunity to the trisection disease, but it is not invariably so. One trisector applied Desargues' theorem in his proof, and that is a theorem that most college mathematics majors do not encounter in their undergraduate years. Another gave a trigonometric proof that was full of partial derivatives. Both of these trisections happened to be more accurate than the general run of constructions, thus illustrating the value of mathematical training in whatever you undertake, even if it is a trisection.

Trisectors have the delusion that the trisection is important. Some seem to think that mathematicians cannot divide angles in three, and that they need to. No one has told them that a protractor will do the job quickly and well. They do not know that Archimedes' straightedge with two scratches on it will also work. An 1892 trisector had picked up an odd idea:

It having been hitherto deemed impossible to geometrically trisect or divide any angle into any number of equal parts, or fractions of parts, the author of the present work has devoted careful study to the solving of the problem so useful and necessary to every branch of science and art, that requires the use of geometry.

Another said

> The study of technical magazines and data shows that a solution is
> being sought whereby a standard construction permits the thrice di-
> vision of any given angle by the simple use of compass and ruler.

but he did not mention the magazines. The most unusual application of the
trisection I have seen was suggested in 1934:

> the TRISECTION OF THE ANGLE may conceivably prove to be
> the key to the discovery of the modern philosopher's stone, by means
> of which it would be possible to change one element into another—
> that is to say, practical alchemy.

One trisector in Ohio refused to send me his construction because he thought
it was worth money, and even though he had copyrighted it he was afraid that
I would steal it. It is very common for trisectors to take out copyrights on their
works or have their pages signed and dated, sometimes with the signatures of
witnesses:

> When the time came for me to submit this project to a publisher,
> I was very much concerned about the copyright. I was fearful that
> if I submitted to a publisher, they might steal the entire trisection
> and I would have to go to court and try to establish my right to the
> trisection.

Never submit to a publisher! Resist always!

Another trisector wanted to know where to apply for the prize money he
had heard was offered for the solution. I do not know where such rumors start.
Nor do I know why there is the persistent delusion that the trisection is important, though it has evidently always existed. For example, here is a Mas-
sachusetts M. D. writing about his trisection in 1890:

> Trusting this effort of mine will prove a benefit to science, I will forgive
> the years of toil for the glory of the achievement.

"I will forgive"; what is more infuriating than the condescension of the
ignorant?

Trisectors draw complicated diagrams. Figure 2.1 (I have left off the letters) is fairly typical. It was too much work for me to reproduce the most extreme example in my collection, but on it the lines and arcs cover what seems
to be all the available space, and it is utterly impenetrable. Its author could
correctly state, "No one has ever found anything wrong with my construction."

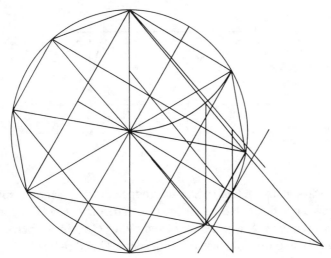

FIGURE **2.1**

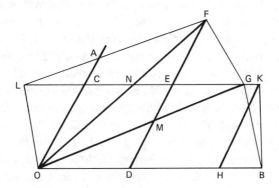

FIGURE **2.2**

Without exception, the constructions can be simplified, sometimes drastically. The reason for the unneeded complexity is not clear. Perhaps the trisectors become so involved with their work that they are not able to look at it with a fresh eye and see how it could be made simpler and easier to understand. Perhaps they do not know that simplicity and clarity are virtues in mathematics. Perhaps they think that complexity is impressive, or perhaps they are a little unsure about their work, consciously or not, and think that errors will be harder to find in a complicated diagram with many letters. Figure 2.2 contains a relatively simple trisector's diagram, but even it can be simplified to Figure 2.3. I

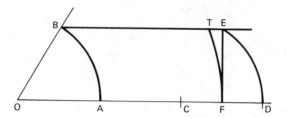

FIGURE 2.3

will not go through the details of how to produce the first diagram, but for the second, draw a line through B parallel to OA, mark off AC and CD each equal to OA, draw the arc DE with center C and radius CD, drop a perpendicular from E to OD, and draw the arc FT with center O and radius OF to get the trisection point T.

As trisections go, this one is not terribly distinguished, and in fact it is a rediscovery of a trisection point which has been discovered by several other trisectors. I include it because it is surely the only trisection ever made by a president of a university. The author, the Very Reverend J. J. C., was in 1931 the president of a third-rank but respectable university. This is more evidence that trisectors are not nuts or loonies, but are perfectly normal, or better than normal, away from their trisections. No person becomes the president of a university, even one of the third rank, without having qualities not had by the common run of people inhabiting universities. Father C. was born in Michigan in 1878, was graduated from the university of which he was to become president in 1897, and was ordained into the Roman Catholic priesthood in 1904. He taught from 1897 to 1912, held a pastorate for five years, and was president of another college from 1917 to 1930. He then moved to the post he held in 1931, retaining it until 1940, when he became pastor of a church in Louisiana until retiring in 1968, at the age of ninety. He died the following year. Besides his trisection, which was only a pamphlet, he was the author of a thick book, *Euclid or Einstein* (1932), in which he proved Euclid's fifth postulate (he thought— proving the postulate is as impossible as trisecting the angle) and refuted, he thought, Einsteinian relativity. There are, by the way, at least as many physics cranks as trisectors, and a large number of them try to demolish Einstein. An inordinate number of them seem to attempt the refutation for religious reasons, as if somehow Newtonian physics has the force of scripture which must be defended from Einstein's undermining efforts. Father C. was also the author of *The Science of Language, The Science of Grammar,* and *Word Study,* books I have never seen but whose titles do not indicate unconventional ideas.

Father C. also duplicated the cube. You would think that he would have consulted his department of mathematics about his work, but many trisectors prefer to work alone and confer with no one, for reasons of fear or pride. Or perhaps he did consult and no one objected. Teaching jobs were hard to come by in 1931.

His construction, which is the one in Figures 2.2 and 2.3, is equivalent to asserting that

$$\sin(\theta/3) = \frac{\sin\theta}{2+\cos\theta}$$

something that could be verified by an intelligent trigonometry student. Father C.'s level of mathematical knowledge was not high, as can be seen from six absolutely incredible trigonometric identities in his pamphlet:

$$\sin(A/3) = 2\sin A + \tan A$$

$$\cos(A/3) = 2\sec A + 1$$

$$\tan(A/3) = 2\sin A$$

$$\sec(A/3) = 2\cos A + 1$$

$$\csc(A/3) = 2\sin A + \cot A$$

$$\cot(A/3) = 2\csc A$$

I am *not* making those up. I *could* not make them up! I could not force my pen to write them! They appear as PROPOSITION XIV, page 18, of *The Trisection of the Angle,* by J. J. C., published by Father C.'s university, dated December 9, 1931, and I am looking at them as I write this. There are no misprints. Father C. had no difficulties with sines and cosines which could be bigger than 1 (or bigger than 10, or 100, . . .). That tangents never exceeded 2 did not bother him. He really thought that

$$\tan(A/3)\cot(A/3) = (2\sin A)(2\cos A).$$

That is, he really believed that

$$1 = 4.$$

No, he did not consult his mathematics department. No teacher of mathematics could possibly have tolerated those atrocities, not even in the depths of the Depression.

Father C. got a lot of publicity. The Associated Press carried stories about him on December 12, 15, 19, and 29, 1931 (at least, and there had been a previous announcement in August), and they generated numerous letters to editors.

In Indianapolis there were several readers who felt called upon to try to set the matter straight, and, in addition, there were two readers who wrote in to say that they had already trisected the angle, one of them five different ways! It is no use to try to put trisectors down in the public press. It may be, as De Morgan said, better to keep quiet:

> When a paradoxer parades capital letters and diagrams which are as good as Newton's to all who know nothing about it, some persons wonder why science does not rise and triturate the whole thing. This is why: all who are fit to read the refutation are satisfied already, and can, if they please, detect the paradoxer for themselves. Those who are not fit to do this would not know the difference between the true answer and the new capitals and diagrams on which the delighted paradoxer would declare that he had crumbled the philosophers, and not they him . . . .
>
> They call mankind to witness that science will not defend itself, though publicly attacked in terms which might sting a pickpocket into standing up for his character; science, in return, allows mankind to witness or not, at pleasure, that it does not defend itself, and yet receives no injury from centuries of assault.   (*Budget,* vol. 2, p. 354.)

On the other hand, we should perhaps continue to wage the war against irrationality, as we continue to fight against ignorance, for fear of the consequences. De Morgan's preceding statement is not definitive, since he also wrote, concerning the circle squarer James Smith ($\pi = 3.125$)

> Why do you take so much trouble to expose such a reasoner as Mr. Smith? I answer as a deceased friend of mine used to answer on like occasions—A man's capacity is no measure of his power to do mischief. Mr. Smith has untiring energy, which does something; self-evident honesty of conviction, which does more; and a long purse, which does most of all. He has made at least ten publications, full of figures few readers can criticize. A great many people are staggered to this extent, that they imagine there must be the indefinite *something* in the mysterious *all this.* They are brought to the point of suspicion that the mathematicians ought not to treat "all this" with such undisguised contempt, at least.   (*Budget,* vol. 2, p. 129.)

The general public may be calmed, but not trisectors, many of whom are beyond the reach of reason. Here is one trisector defending himself:

> Those who are skeptical should offer something more than rhetoric or argument in order to disprove geometrical *facts.* Assuming the angle

and its trisectors given, the enveloping quadrantal arc constructed, and its points of trisection found, if it be denied that the trisectors pass through these points of equal division on the quadrantal arc, let them show *by the ruler and compasses* where these lines and points *are* with respect to each other on the quadrant. If the lines constituting the respective pairs of trisectors of both sectors do not intersect on the quadrantal arc they should show by *the ruler and compasses where they do intersect.*

That is at least comprehensible, unlike some trisectors' writing, in which the meaning is undiscoverable. But it is not logical: the author is saying that if you want to prove him wrong, you have to produce a trisection with straight-edge and compass. Trisectors will not listen to reason. Nor will circle-squarers, as De Morgan on James Smith illustrates:

Mr. Smith's method of proving that every circle is 3 1/8 diameters is to assume that it is so,—"if you dislike the term datum, then, by hypothesis, let 8 circumferences be exactly equal to 25 diameters,"— and then to show that every other supposition is thereby made absurd.

"I think you will not dare to dispute my right to this hypothesis, when I can prove by means of it that every other value of π will lead to the grossest absurdities; unless indeed, you are prepared to dispute the right of Euclid hypothetically for the purpose of a *'reductio ad absurdum'* demonstration, in pure geometry." (*Budget,* vol. 2, p. 117.)

Using the same reasoning, I must be a billionaire because, if I make that assumption, then any other conclusion about my net worth is made absurd.

Trisectors are great letter writers. Mostly, they correspond eagerly with any mathematician they can and are hard to get rid of. An extreme example is a crank, not a trisector, who made the following discovery. Take the six permutations of 1, 2, 3; arrange them in increasing order; and write the differences between successive terms:

123		132		213		231		312		321
	9		81		18		81		9	

The sums of the two lines are 1,332 and 198. Now, take the first nine digits of ($\pi = 3.14159265358979\ldots$) in groups of three and add that magic constant 198:

$$
\begin{array}{ccc}
314 & 159 & 265 \\
+198 & +198 & +198 \\
\hline
512 & 357 & 463.
\end{array}
$$

Though it is tempting to consider the numerological significance of $512 = 2^9$ and 357, with odd digits in increasing order, we will continue, as did the crank. The sum of the three totals is

$$512 + 357 + 463 = 1,332,$$

an astonishing coincidence, one whose chance of happening is on the order of 1 in 1,332. I wrote to the crank to tell him so. In the next three weeks I got *twelve* multipage letters, even though I replied to none of them, and later he gave it a few more tries. Things do not change: here is De Morgan on his $\pi = 3.125$ circle-squarer:

> Mr. Smith has written me notes: the portion which I have preserved—I suppose several have been mislaid—makes a hundred and seven pages of note-paper, closely written. To all this I have not answered one word, but I think I cannot have read fewer than forty pages. In the last letter the writer informs me that he will not write at greater length until I have given him an answer, according to the "rules of good society." Did I not know that for every inch I wrote back he would return an ell?   (*Budget,* vol. 2, p. 123.)

It is almost always a mistake to correspond with trisectors because it is virtually impossible to convince them that they have made errors. The circle-squarer James Smith wrote to De Morgan

> You may as well knock your head against a stone wall to improve your intellect as attempt to controvert my proofs.

De Morgan wisely wrote

> I thought so too; and tried neither.   (*Budget,* vol. 2, p. 245.)

One trisector sent me copies of correspondence he had had with a mathematician with the Rand Corporation. No doubt the mathematician had fun when writing

> I thought you would be interested in my method of angle trisection. Since the construction is rather simple (much more so than yours) and involves no change in the compass setting, I thought it would have mystic appeal for you.

He carried on in a heavily ironic vein. He paid a price for his fun, because the correspondence continued for at least *seven* years. It is best not to start such things. A circle-squarer, around 1750, bothered the mathematicians Johann Bernoulli and Samuel König with his material. Bernoulli replied

Following the hypotheses in this work, it is so evident that $t = 34$, $y = 1$, and $z = 1$ that there is no need of proof or authority for it to be recognized by everyone.

Bernoulli, as the Rand mathematician, was having a little fun. König was, also:

I subscribe to the judgement of M. Bernoulli as a consequence of these hypotheses.

A little fun, and in addition the circle-square went away. But look what he wrote later:

It clearly appears from my present Analysis and Demonstration that they have already recognized and perfectly agreed that the quadrature of the circle is mathematically demonstrated.

(*Budget,* vol. 1, p. 150.)

You cannot win, at least not often, no matter what you do.

A consequence of the love of letter writing is that trisectors as a group can take a great deal of time from the mathematical community, not to mention money for postage. Most mathematics departments do not bother responding to crank letters, but some teachers of mathematics are so filled with the urge to educate that they try to reason with the trisector. This is almost always futile. You may come close, but you will seldom succeed. A 1975 trisector in Guyana wrote

One great scholar wrote, and I quote: "I feel I should tell you that it was proved a long time ago that it was impossible to trisect an angle using ruler and compass alone: consequently, any attempt to perform this construction is bound to be a waste of time."

He then prepared for me a simplified thesis proving the impossibility; and this nearly shattered my faith in the Supreme Mathematician, to whom I had made intercession for revelation. However, intuition and grace kept me constant in the face of many reverses till revelation came, pointing out to me that the requirement for the trisection of angles was the construction of a fourth root and not a cube root as the scholar had labored to prove.

Labor in vain—proof is helpless against revelation!

Some trisectors waste vast amounts of mathematical time. A 1951 trisector in Michigan broadcast his construction to the leading state university of each state, to prestigious private institutions, to Albert Einstein, to over two hundred places in all. He had more than sixty replies! Think how many mathematician-hours went into their preparation. The trisector excerpted ten or so of them in his next widely distributed letter, which showed that their contents had no effect at all on his belief in his trisection:

> From the National Academy of Sciences: "There is no issue. The question has been settled once and for all."
>
> From *Mathematics Magazine*: "This has long ago been proved impossible with ungraduated ruler and compass alone."
>
> From Chicago: Chicago will examine solutions for a fee, to cover the time cost.

There were other answers from MIT, Columbia, Cornell, Illinois, but the best was Einstein's, which used the delightful and effective formula

> I am so overwhelmed with correspondence that despite every desire to do so I have no time to reply to all my letters.

Some trisectors have the means to publish and distribute their works widely; this has two bad effects. First, the trisector is wasting resources which could be put to better use, and second, there is the danger that someone may read and become convinced, or at least confused. One of De Morgan's circlesquarers—Milan, 1855, $\pi = 3.2$—also had many replies:

> [The circle-squarer] is active and able, with nothing wrong with him except his paradoxes. In the second tract named he has given the testimonials of crowned heads and ministers, etc. as follows. Louis Napoleon gives thanks. The minister at Turin refers it to the Academy of Sciences and hopes so much labor will be judged worthy of esteem. The Vice-Chancellor of Oxford—a blunt Englishman—begs to say that the University has never proposed the problem, as some affirm. The Prince Regent of Baden has received the work with lively interest. The Academy of Vienna is not in a position to enter into the question. The Academy of Turin offers the most *distinct* thanks. The Academy della Crusca attends only to literature, but gives thanks. The Queen of Spain has received the work with the highest appreciation. The University of Salamanca gives infinite thanks, and feels true satisfaction in having the book. Lord Palmerston gives thanks. The Viceroy of Egypt, not yet being up in Italian, will spend his first moments of leisure in studying the book, when it shall have been translated into French: in the mean time he congratulates the author upon his victory over a problem so long held insoluble. All this is seriously published as a rate in aid of demonstration. If those royal compliments cannot make the circumference about 2 per cent. larger than

geometry will have it—which is all that is wanted—no wonder that thrones are shaky.   (*Budget,* vol. 2, pp. 61–62.)

One trisector had his construction published as a hardbound book by a vanity press, complete with a picture of the author, a pleasant-looking old gentleman, on the back cover. I must have been in a harsh mood when I concluded a letter to him with

> It is a shame that you have spent so much time, energy, and money trying to do the same thing as trying to prove that the final score of a football game could be 7 to 1; that is impossible, and it can be proved. It is also a shame that you have seen fit to publish a book which can only mislead its readers and spread error.

I swing from sympathy to exasperation with trisectors. I do not presume to tell experts in other fields how to do their business; why should I be patient with people who do that to me? On the day I wrote to him I must have been especially exasperated. I was properly shamed by his reply, which said in part

> I am most grateful to you for the interest you are showing in my tri-section, and the time you have taken to correspond with me.

Pictures do not lie: he *was* a nice old gentleman.

Now, will you know a trisector when you see one coming? And will you know what to do? Here is a hint: what you do involves your legs. No, you do not kick him.

Three Trisectors

There is one trisection which is of more importance than that of the angle. It is easy to get half the paper on which you write for a margin; or a quarter; but very troublesome to get a third. Show us how, easily and certainly, to fold the paper into three, and you will be a real benefactor to society. (*Budget,* vol. 2, p. 15.)

Nothing worse will ever happen to me than the smile which individuals bestow on a man who does not *groove.* Wisdom, like religion, belongs to majorities. (*Budget,* vol. 1, p. 30.)

One summer I decided to visit some of the trisectors with whom I had been exchanging letters, to see what sort of people they were. This was in a way exploiting them, since there was nothing they or I could do face to face that had not already been done by mail, but none turned me down and told me to stay away. On the contrary, they would all be most happy to see me, they said. There follow accounts of my visits to three trisectors. I think that these three represent a very large percentage—I am tempted to say 100 percent—of the types of people who trisect.

A. B. was the first of the trisectors I visited. He lived in a southwestern state in a new subdivision built around a manmade lake, one of whose purposes was to provide water for a large city fifty miles away. The subdivision—all of whose street names were Hawaiian or related to Hawaii—had fewer than half of its lots occupied. The lots with buildings on them had houses of all sizes, some permanent residences and others clearly for weekend or vacation visits, and many were occupied by mobile homes. The asphalt in the streets was cracked in places, and there had been frost damage.

I found A.'s house with no trouble, and he greeted me with great enthusi-
asm and hospitality. He was ready for me to stay days and days, though I only
spent one night there. He lived with his wife in a six-year-old house; it had
only one bedroom and was not large. A. was about six feet tall, of a medium
build tending toward slenderness, and had thinning hair. He was fifty-three
years old when I met him. His three sons had grown and left home. His wife
of twenty-nine years had a pronounced southern accent, much more so than
A., whose pronunciation hardly differed from standard American. Before talking any mathematics, I had to be shown the tree limb in the bathroom. There
had been a storm the day before and a tree had fallen on the house, breaking
through the roof.

A. had a great diversity of interests. Some of his paintings, large abstract-
impressionistic works, decorated the house. He had written two novels, one
published by a southern publisher I had never heard of. He had not had good
luck with it:

> My publisher—now this is my bad luck—my publisher got killed two
> months after that book came out. He was just in the routine of going
> up to all these TV stations all over the country getting everything set
> up for interviews and all that kind of thing. He got killed in an auto-
> mobile wreck. Another guy down in [a southern state] took the novel
> over and it hasn't done worth a dern since then. I've sold forty-five
> copies out of the three thousand he had printed up.
>
> Who cares? I know it's a good book, he said it's a good book, the
> guy down in [the southern state] says it's a dern good book. But it's
> promotion that sells a book. That's what goes today, promotion. It
> doesn't go on its merit, it goes on the ballyhoo that's out there in
> front of it. Not right. But I can't worry about that.

His book is subtitled *A Diary of Deliberate Death,* and it is the diary of a man
who starves himself to death. It is illustrated with ink sketches by the author and
is curious in not having one comma in it anywhere. Instead, A. used a slash (/),
and I was surprised to find that it was not an annoying affectation. The slash is
not always used for a comma/sometimes it replaces other signs of punctuation.
Here is a characteristic excerpt from the book:

> From starvation to satiety in twenty-four hours. Let me see. Vainglo-
> rious! Vainglorious! So you're Ronald Firbank/I presume! A beery
> smell/the tinkling piano/muffled voices/eyeballs out of a smoke-
> fog/ leopard-skin panties/and a suggestion of apes/pearls and the color
> blue. Green door/green door/under red neon . . . open to my knuck-

led knocker! "Sir/let me tell your fortune/" a mauve-lipped black-amoor/fruitbowl on her haid/spoke.

To my critically untrained eye, this is every bit as good as some things I have read in collections of experimental writing.

A. was a wide reader. He had bought copies of several mathematical works: Edna Kramer's *The Nature and Growth of Modern Mathematics* (a large and thorough popularization) and other standard books meant for the general public. More than one of them had mentioned the trisection and its impossibility, but they had no influence on A.'s thought:

> "Wantzel was the first to give a rigorous proof of the impossibility of trisecting the general angle by straightedge and compasses." I think I read that the other day in that book I'm reading in there. But what proof does he really offer that it can't be done?

There was difficulty here with the meaning of proof. More came out later, when A. emphasized that his construction used the Pythagorean theorem and another theorem of geometry:

> I've often heard that if you use back theorems that have been proven, you can prove it through these .... OK, but I've used these two theorems that are already well-known and proved many thousands of years ago that check out in my trisection. Why doesn't it prove my trisection is right there?

It seems that what A. was saying was that if his construction did not *conflict* with something true, then it must *be* true. If this is a common way of thinking among trisectors, it partly explains why proofs of impossibility do not impress them. A. was also not impressed by my repeated assertion that his trisection was only an approximation. He said that bisections are approximations too and are probably less accurate than to five decimal places (the accuracy of A.'s trisection). He meant bisections with physical instruments of course and he concluded that the error in either construction was of the same order of magnitude. That is true, but his further conclusion that bisections and trisections are on the same level and that his trisection is as good as a geometry book's bisection does not follow. I did not find this argument an easy one to answer, not that any answer would have made any difference. It must not be thought A. was unable to reason logically. He could, as an example I'll give shortly shows.

After he discovered his trisection he had a booklet printed:

> I got busy with my wife, we had these things printed up, we put them in the bedroom there in our apartment and put all those things together,

four hundred of them. Stapled them together, that was a lot of work, I'll tell you, but I enjoyed doing it. I sent it out to three hundred—well, not three hundred, I sent a lot of universities three copies and some of them two and some one, according to what I thought they'd use. I went down to the library to the book that had all the different colleges and universities in the country with addresses and zip codes, and just sent them off at random. I didn't know anything, I just sent them to the professor of mathematics, mathematics department, so-and-so university.

He got a number of replies, which led to his purchase and reading of mathematical books. He showed me some letters from publishers, declining to publish his work, though one had gone so far as to send it to a university mathematics department for an opinion. One of the replies was from a mathematician who for some reason had enclosed an offprint of a paper of his own on coloring spheres in which he showed that a certain problem had a five-color solution but could not be solved with three colors. He must have had an oversupply of copies. A. responded with an erroneous refutation of the paper to which the mathematician did not respond in spite of several later letters from A. I saw where A.'s refutation was wrong, showed it to him, and he understood it. He said, "Boy, [he] must think I'm pretty dumb. I bet that's why he didn't answer." It's likely that the mathematician would not have answered in any event. But A. could reason: he understood why his refutation was not right, though he would not understand that his trisection was not right either.

It checks out every time. It *has* to be. These lines going out, what do you call them, wings of this angle I've already trisected. Why doesn't that prove this right?

Talk did not help. Here is part of our conversation to show, a little, how the minds of trisectors (and of their wives) work.

A.:   A mathematician is just a human being like I am.

Me: Yes, but a mathematical theorem is independent of humanity after it is proved. It is true, once and for all.

A.:   But who started this geometry then? I mean, who actually did it?

Me: A person did it.

A.:   A human being.

Me: Yes, but it will never need revision or changing. This is how mathematics differs from . . .

Mrs. B.:   Yes, but look . . .

A.:   It's so beautiful, boy . . .

Mrs. B.:   Well, all right, somebody has to do it in the first place.

Me: Right.

Mrs. B.:   OK, but who's to say that this isn't the first one to do this particular trisection and be right?

Me: The difficulty is that it conflicts with . . .

A.:   What thousands of mathematicians have said before and they're all human beings. It's all based on π. Let's junk this π and start this mathematics thing all over again.

Mrs. B.:   Oh, fiddle. In order words if it doesn't have any . . .

Me: I know I'm not going to convince you.

A.:   But you're not going to make me mad either. See, I'm not going to get mad. (Laughs.)

Mrs. B. No, he doesn't. It's all right, he can take . . .

A.:   I can take anything you say.

Mrs. B.:   He takes it better than most people.

A.:   I want to discuss it with you; I'm anxious to discuss it with you.

Me: Here's the Pythagorean theorem, for example.

A.:   Which I think is something really wonderful.

Me: And a man did it, but it's . . .

A.:   Irrefutable.

Me: Yes. In chemistry, the laws may need to be changed; the laws of physics may need to be changed; in fact they were: Newton discovered these laws that had to be changed by Einstein, but . . .

A.:   Wait a minute, there.

Mrs. B.:   Just a minute, let him finish.

A.:   Let's stop right there. OK?

Me: Yes.

A.:   A lot of things in mathematics have changed too, over the years.

Me: No. Once a theorem is proved, it does not change. There are fashions in mathematics, and subjects come and go . . .

A.:   But why do I keep reading in mathematics books, there's a great big thing I'm reading just now, that there have always been changes made in mathematics right along. Euclid has been proven wrong in lots of cases.

Me: Yes, in the past standards of proof were not what they are today, and it wasn't insisted on that the logic be perfect in every detail. But then . . .

Mrs. B.:   It all boils down to the fact that mathematicians say that this has never done; it can never be done; so regardless, it's not done now.

Me: Yes, they're saying it can never be done because it's been *proved* that it can never be done. There's this theorem, just like the Pythagorean theorem, that says . . .

Mrs. B.:   That you just cannot trisect an angle. I mean, that's it.

Me: Right. No matter how hard you try.

Mrs. B.:   And they're willing to just let it go at that and say, this is it? Just to close the door and that's it.

Me: Yes, because if this theorem is correct, then any trisection must have a mistake. So, really, if anyone is going to say, I can trisect the angle, the first thing they should do is show where the flaw is in this theorem that says it can't be done.

Mrs. B.:   Oh, I see, I see.

A.:   What theorem is this you're talking about? That says it can't be done?

Me: Well, it's outlined in Yates's book . . .

A.:   In this book you gave me?

Me: Yes.

A.:   Am I supposed to keep this? Do you have another copy?

Me: I can get one.

A.:   It gives all kinds of approximations in here. Mine doesn't fit any of the ones he's ever seen. He says mine's probably a whole lot the most simple he's ever seen, really.

Me: Yes, it's very neat. Elegant—a technical term.

A.:   And I was very enthused when I got through working on it. I still feel enthused about it.

Mrs. B.:   I don't blame you. I would too, regardless.

Me: Mathematics is fun.

Mrs. B.:   You can't help but be proud of something if you feel like you've done something and ...

A.:   With nobody else's help.

Mrs. B.:   That's exactly right.

Me: It's a remarkable achievement.

Mrs. B.:   Certainly it is. That's right.

A.:   Nobody else has ever done it. I mean, they've done it different ways, but I didn't do it the way they did it. There must be, how many thousand ways are there to do the dern thing?

Mrs. B.:   Probably thousands of people have actually trisected angles before.

Me: They've *tried* to.

A.:   I keep asking, why hasn't someone else come up with this simple solution that I came up with?

Mrs. B.:   Because of this right here, and it won't be accepted; this is the reason right there.

A.:   At least I can be congratulated on my elegance.

Me: And originality, certainly.

A.:   [Laughs.]

Mrs. B.:   Well, though men have trisected angles before ...

A.:   Approximately.

Me: Yes, using extra tools.

A.:   I told you, Archimedes did it, honey, but he cheated.

Mrs. B.:   Well, I mean, somebody 50 years ago trisected an angle, or 25 years ago.

[The ghost of Father C.?]

Me: All you have to do is get your protractor out.

Mrs. B.:   Except he never used a protractor. He used a straightedge and compass; that's all he ever used.

A.:   Yes, I threw away my protractor. I never used it.

Mrs. B.:   Yes, that's right; he didn't.

A.:   I'm not interested in degrees; it's the trisection every time. I tossed that aside completely when I started the problem. I didn't worry about degrees; I just worried about the trisection of a drawn angle of unknown degrees.

Mrs. B.:   Yes, that's right.

A.:   I want to try to explain. Archimedes took a compass and straightedge that he made a mark on, which is not allowed; you can't do that. He put a mark on it with a pencil or something. He drew his half circle and marked it off in some certain way; I can't explain it to you, but anyway he did it that way. That's not fair. But he was a pretty intelligent man; don't get me wrong. [Laughs.]

Mrs. B.:   Well, anyway, I'm proud of him and he's proud of himself. I said he's never going to get any recognition for it, and he said he probably wouldn't, but I said, if you think you've done it, this is fine.

The conversation was a long one, both before and after this portion. I later made the mistake of mentioning to A. another problem that could be solved only approximately with straightedge and compass, that of squaring the circle. After I left, he started to work on it, in spite of my clear statement that an exact construction was impossible, had been proved to be impossible, and many people had wasted a lot of time in trying it. In about three weeks I had a letter from him with a construction which, he asserted without proof, squared the circle. It was a crude approximation which was easy to refute because A. had drawn it on graph paper: just count the squares, I told him, in the circle and in the square, and you will see that the difference is too large to be caused by inaccurate drawing. That, of course, did not take care of that. Later came another construction, this one so much improved that square counting was not enough to show the error. Out had to come the trigonometry to show that the construction was equivalent to saying that π was $(16/9)^2$ which it is not ($3.14159\ldots \neq 3.16049\ldots$). Later, another more complicated construction came—all of these with no attempts at proof—and I chose to give up.

I gave A. the names and addresses of two other trisectors. Most trisectors do not know that there are others like them, and knowing that they are not unique and that others have failed might influence them to give the trisection up, I thought. I was not acting entirely in A.'s interest, but in self-interest too: one of the trisectors was an old man in Ohio who refused to let me look at his construction because, I think, he was afraid that I would steal it and get all of the money and glory, even though his construction had been copyrighted, and the other was a trisector who sent me long and frequent letters, most not capable of being answered, and I hoped to give him another outlet for his energies. Acting selfishly is not good, and I was properly served by not having anything happen that I wanted to happen. The old trisector did not let his secret out, the letter writing trisector did not quit writing, and A. did not lose his faith in his trisection.

At that time, A. had no job, though he had been in the navy for a number of years. He had gone to a school for radio announcers, he said, and had for a time a job with a radio station, but it had not lasted long. Nevertheless, he was cheerful, ebullient even, and full of energy. He later moved to a western state, but after a short time moved again to a small town in the southwest. In his last letter, telling me of his moves "away from the tornadoes and the crazies," he said

I have now trisected the angle, squared the circle, and doubled the cube . . . . Now what can I work on next?

Since he had the ability not to take himself too seriously, I think that the trisection was only an episode in A.'s life.

For the next trisector, the trisection was not only an episode. When I met him, C. D. lived in a small city in the midwest, home of a state university. Unlike many trisectors, D. was a man of education. He had a B. S. degree in chemistry from a good liberal arts college and an M. S. degree in the same subject from a large state university. He had, he said, studied for a doctorate, but had never completed the requirements. He was a teacher of chemistry and physics in high schools for many years, but was no longer, because of the trisection.

I took mathematics in high school and I think my teacher suggested there was a problem like this. I think maybe I had just a momentary interest in it, but in 1937 I was teaching and I didn't have any money to go anywhere Thanksgiving vacation. I tried something without looking up the historical efforts . . . it took me about three days to do this, and so the vacation passed. I didn't pursue it to any degree until 1963.

There were early indications that D. might turn out to be a trisector: a professor in the chemistry department where he had pursued his doctoral studies said

He [D.] says his graduate research director didn't understand the project he wanted to carry out, which included many of the seeds of a theory of bonding that anticipated the work which won Linus Pauling the 1954 Nobel Prize in Chemistry.

Another professor had told the chemistry professor that

Mr. [D.] was attempting to disprove Einstein's theory of relativity during his graduate days. If I remember correctly, he recalled that Mr. [D.]'s research director just wasn't able to work with him and that finally [D.] was judged not a competent Ph. D. candidate and so was dropped from the department.

It was in 1965 that D. achieved his construction and started on his career of trying to have it accepted. He did this for many years. Since our correspondence terminated many years ago, for reasons which will be mentioned, I do not know how long he kept it up. His efforts have included sending his trisection to several mathematics journals, of course with invariable rejection; trying to speak at scientific gatherings; and trying to interest professors at the state university in it, not only professors of mathematics, but of chemistry, physics, and engineering. In 1969 he was able to address a group from the physics department of the state university:

[The talk] attracted a score or two of people, largely high school students here on an NSF summer program .... [D.] and the audience failed to communicate. This was partly because of [D.]'s rapid presentation in which he assumed the audience was much better acquainted with his construction and the history of the trisection problem than they were. Also, he was not able to grasp the nature of the questions asked of him by the few real mathematicians present.

His talk was written up in the local newspaper. The reporter may have thought he was being "objective" by not mentioning that the trisection was impossible:

A possibly revolutionary theory in mathematics was offered ... yesterday .... In a question and answer period, he was asked what specific objections have come up as to why the scientific world hadn't accepted his idea.

"None, I would say, except the fear of examination. Over 250 people trained in mathematics have looked at it with no objections. Three professors at [the state university] promised to give me an evaluation. So far, no mistakes have been pointed out, but also, no evaluation," [D.] said.

The professor of chemistry quoted previously went on,

He has been trying since 1965 or 1966 to present his work at the [state scientific organization], so that it would appear in the published *Proceedings*. Various feature articles have appeared from time to time in newspapers around the area as Mr. [D.] has gone from one teaching job to another, convincing small town officials of the importance of his work. Apparently at its May meeting this year, the [state scientific organization] granted him a room and a lecture time but did not list him on the official program.

A professor of mathematics at the state university was approached by D. What happened was classical:

In 1970 I agreed to evaluate Mr. [D.]'s work, and during 1970–71 met with him three or four times and we exchanged correspondence rather frequently. At the time he initially approached me he seemed quite friendly and sincere. I felt that perhaps he had been an unfortunate victim of harsh treatment in the past and I found it easy to sympathize with his desire for an evaluation.

Actually, trisectors seldom seem to get any harsh treatment. Some mathematicians ignore them, and others, in gratitude for being asked their opinion (which does not happen to mathematicians very often), in effect lead them further down the slope. Harsh treatment may be what trisectors need.

> Things went quite well at first. His "proof" of his trisection was crude to the point of being unintelligible but he began to clarify things with a series of revisions. Of course I invariably found a flaw and he became increasingly frustrated.

Why would a mathematician allow a trisector to continue a search for an impossible "proof"? Probably because at each face-to-face meeting, the easiest thing was to point out a new flaw, hoping that D. would not be able to get around it and would thus never come back. But the persistence of some trisectors goes beyond any limit that a mathematician can understand. Also, it becomes second nature for teachers to encourage students to try to master material when they know that in all probability they will not. It would be very hard for a true teacher to say, "Your proof is no good, it will never be any good, nothing else you try will be any good, it is no use, stop trying, and go away."

> He refused to study modern expositions of the "disproof" or even seriously consider the possibility that his trisection was invalid. Then he challenged me to open debate. When I refused he accused me of intellectual dishonesty and of conspiracy to deprive him of his rightful recognition. When he finally accused me of causing him to lose his job I had to inform him of my desire to terminate all further contact with him. He made only one further attempt to see me but that was a feeble one. He did, however, register complaints with the University President and with the American Association of University Professors (!)

The professor of chemistry quoted previously shows how D. went from construction, to proof, to refutation of Wantzel. At each stage, the path taken is the one of least resistance, but it leads in the end to maximum harm to the trisector. A chemist, however, can no more be expected to have the details of the trisection at his fingertips than a mathematician can be expected to be up on the valance of phosphorus.

> At first he almost had me convinced that his approach bypassed implicit assumptions in the proof that it couldn't be done. However, after I read this proof and discussed it with some of our math faculty, it became clear to me that Mr. [D.] did not understand it at all.

> In fact, at first, Mr. [D.] was convinced that by carefully carrying out his construction on a large scale he could show visually he had indeed trisected the angle and therefore no other proof was necessary.

I finally convinced him that the proof of a construction was the only thing that would interest anyone.

The first few proofs he developed amounted to assuming he had established the valid trisection points in the angle and then merely deriving some of the properties of such points. Over the years his proofs became more complex and I lost interest early in trying to find holes in them.

That was sensible, because the holes would have been repaired and replaced with other holes that would have been harder still to find.

Some of our math faculty put up with him until he became unreasonable and, they thought, abusive. In recent years he has become increasingly bitter and depressed because he can't arouse the interest of editors of mathematical journals or mathematicians generally, or even get them to correspond with him, as I'm sure he told you. Consequently his attempts to arouse interest have taken increasingly bizarre and provocative forms, which have almost led to legal actions . . . .

Mr. [D.] has suffered considerably from his trisection efforts and his unfortunate belief in their importance.

When I met him, C. D. lived in a small house in an area where the houses looked very much alike, about 20 years old, surrounded by grass and set on gently curving streets. The trees had not yet grown enough to overcome the impression of newness and openness. The D. house was crowded with furniture, perhaps as a result of a move from a larger house. There were bookshelves with a variety of books and magazines, including a number of religious works. There were three Siamese cats. Mrs. D. was a plump woman, in her late fifties I thought, with unlined skin with a powdery look to it, gray-white hair, and bright blue eyes. Her speech was that of an educated person, and I guessed that she and D. had met in college. She stayed in the room during our discussion, but she took no part in it and after a time she dozed off.

D. was prepared for me with a table furnished with various relevant papers, a ruler, compass, slide rule, and a library copy of Yates' *The Trisection Problem*. He was a man of middle height, bald except on the fringes, not fat, and sixtyish; he in fact looked like a high school science teacher. He talked without prompting, continuously, and he did not seem to listen to me when I tried to say something. His speech was vehement and emphatic, at least on the subject of the trisection, and it matched his harsh and angular handwriting, with which l was already acquainted. Almost all of what he had to say was on his construction and why the proof that the trisection was impossible is not right.

He had five reasons for this, and since they were not to the point, it was not easy to make a satisfactory response to them. But I thought that no response I could make could be satisfactory, so I did not press any point. D. referred to his early efforts:

> Since I hadn't any geometrical proof at that time, since I didn't have any analytical equation, since I merely had something that looked good, I knew it was not satisfactory or sufficient. So, I acted the way most people do when they get the bug of trisection. Whenever they start it, they seem to be infected with something that prevents them from stopping; it is the interest, or curiosity about the problem, or the efforts to reach a solution. Some people, as far as history is concerned, almost removed themselves from reality and became all wrapped up with trying to solve this problem, without success.

What clear vision for others! What blindness toward himself!

> I have copyright for two methods and I maintain there are two other methods if anyone wishes to work them out. Maybe one person already has. My wife and I both attended a meeting in Chicago and I talked to someone there who was from Washington, D. C., and lo and behold! four months later this description appeared in *Mechanix Illustrated.*

That was the 1966 trisection by H. C. which appears in the Budget of Trisections.

> [It] was according to my description a rectangle-and-three-circles method. I wrote to both the author and the editor of the magazine and received no answer. But I maintain that I had indicated my method to someone else and consequently I had prior claim to the idea.

He started sending his construction many places and asking people to evaluate it. He maintained that no one was able to find an error. It is possible that many of those to whom he sent his work did not bother to look at it, and it is possible that many who did look at it did not have the training needed to be able to find an error, and the fact that they did not respond pointing out errors may have been enough to convince D. that no error existed. His efforts had an unfortunate consequence:

> But I guess as a result of asking a lot of people, I incurred somewhat of an antipathy from some of those who were my superiors and supervisors, and as a result, they considered that if I were able to do something which the scientists and mathematicians thought was impossible, maybe I was in the wrong field. So, consequently, to make a long

story short, I went fishing in a different pond, by request, you might say. As a result, I have been doing work of a rather menial sort for several years. I was even unemployed for a while. I have done some selling—one week I sold four welders and got a $50 fan for a bonus and beat somebody who had been selling them for fifteen years—but that didn't last.

When I saw him, D. said he was working as a janitor in two places, 80 hours a week, and was having trouble keeping going. He had written to me earlier

Since I have lost my professional position due to unfair discrimination and failure to have promised evaluations by qualified mathematicians, my children had no financial help and were unable to complete college work for degrees.

He was not a happy man.

According to D., one mathematician, unable or unwilling to try and find any more errors in D.'s proofs, said

Why don't you try to change your tactics a little, though, and do this. Look at the classical disproof and point out where the errors are in it. And if you do that, then you will destroy what the mathematician believes and as a result then maybe he will be a little more friendly about listening to what you have to say.

This was really irresponsible, even cruel, behavior by the mathematician. It was the easiest way to get rid of D. at the time, but the mathematician must have known that D. would be unable even to understand the proof, much less find any errors in it, especially since there are none. D. succeeded in finding what he thought were errors in the proof. The version he saw was a proof by contradiction: "Suppose the angle θ can be trisected," it starts out, and ends by deriving from that assumption something that is false. Since it is impossible to derive false conclusions from true assumptions, it follows that the angle θ cannot be trisected. The proof contained a diagram showing the trisected θ. D. maintained that since the diagram could, according to the author of the proof, not be constructed by Euclidean means, conclusions about Euclidean geometry made from it could not be valid. I tried to explain that the way the diagram was constructed was immaterial to the proof, but I do not think D. was listening.

Later on, he showed me a diagram.

Here, for instance, if you take the angles from 0 to 90 degrees, this is the angle and this is the value of the chord of the angle, if you look along all these little pieces, this would be a linear relationship.

He was constructing a table of chords, even as had been done by Hipparchus of Nicea, under the bright and pristine Mediterranean sky, more than 2,000 years ago. He was rediscovering the sine function. He was reinventing trigonometry.

If you look there, you will find some variations which are obvious; instead of being just a smooth variation, there are some irregular variations.

They were bumps caused by inaccurate measurements. I tried to tell him that it was known that the graph had no bumps, but it did no good:

I think you will find slight variations.

I had brought my computer calculations which showed the magnitude of the error very small in his construction.

Yes, I quarrel with the computer results, for one reason. This is what I was going to say a while ago, that as far as the proper theory is concerned, if you take a series of cases and program them on a computer, you will get an average which may or may not be one of those specific cases. But if mine is correct, with proper theory you will get an exact value which is one of those cases regardless. It will not be an approximate value, even on a computer, because Euclidean geometry is either right or wrong.

I could not see what was behind this.

Toward the end of our conversation D. claimed to have succeeded in squaring the circle and duplicating the cube, and he gave hints about having solved some other large problems, but he had not made his results public. He said that sometimes he felt like burning them. He thought that the trisection might have applications to physics, and he was very unhappy that the advance of human knowledge had been held up by the mathematicians with whom he had been in contact. He called the proof of impossibility a great hoax. I did not ask what the point of such a hoax would be. After I left and was in my car, he came quickly out of his house, perhaps at the prompting of his wife, to give me directions to the main road. They were really not necessary, since it was only two blocks away.

In the week after my visit, I had three letters from D., saying that I had not offered any effective arguments against his position and challenging me to answer his five points about the proof of impossibility. Since the points

the diagram is illogical,
the diagram is plagiarized from Archimedes,
the diagram is non-Euclidean,

the trigonometric solution of $\cos 3\theta$ is not a general solution,

Galois' one-circle presumed explanation is incomplete as it does not use Appollonius' three-circle method,

cannot be answered to someone who does not fully grasp the idea of proof by contradiction, D. got one more proof of their unanswerability from my failure to answer them. I think what D. was doing was *regaining his equilibrium*; I had given him an unfavorable evaluation, I had pointed out an error in his proof, but I had somehow to become number 251 on the list of mathematicians who had been unable to refute his trisection. I am sure he was successful, for he was obsessed and beyond reaching.

Letters continued for some months more, growing more and more verbally violent and abusive. I replied to a few, in one offering sympathy for the hardness of his lot. That was not what he wanted. His reply was written with a blue felt-tip marker on a legal-size sheet of paper in huge letters. Part of it was

Stop commiserating me. You have *not* sent your calculation. Why? Don't I deserve fair treatment? Or are you afraid to have your work verified? You have sent no analysis of errors in my logic. Do you accept my destruction of the classical disproof?

The letters were now more than an inch in height.

You have sent no evidence to the contrary. I use *reality*. What do you use? No answer means you resign to my evidence.

When he was writing it he must have been seething with anger, consumed with frustration, wild with pain. After a short time, repeating a previous pattern, he wrote to the president of my school, abusing me and accusing me among other things of dishonesty. The president passed the letter to the dean, and the dean sent me a copy of his deanly reply, which concluded

Since mathematics is not my field, I do not follow the implications of your letter, but I am sending it on to Professor Dudley. I am sure you will hear from him.

He did. The correspondence had long been fruitless for both of us, so I wrote a sharp note, mentioning the laws of libel. It had its effect, and I have not heard from D. since. I am sure that he continued on, repeating the same pattern. I am sure because of the content of his letters: they were all essentially the same. Written at different times and in response to different things, they all were the same: the same arguments, the same points, the same phrases, over and over again. The channels in D.'s brain were so deep that he was trapped in them, and there was no hope of change. Being too old to start anything new even if he wanted to and having invested too much of himself and his life in his trisection, he was doomed to disappointment and continual frustration until death. Was he tragic? Or illustrative of a lesson? Or was he only one of the millions of unknown casualties of life?

When I met E. F., my third trisector, he was seventy-eight years old. Before and after we met we exchanged many letters. His were written in a curious style, simultaneously formal, polysyllabic, and filled with idiosyncratic punctuation. His first letter to me gives the flavor:

> Dr. [ ] has just informed me of a much appreciated favor that he has rendered me in supplying you with a copy of my trisection of an angle and proof of same. To find someone actually asking for such a thing is a little more than I could have hoped for. I am very glad that you did. I suppose that you would naturally expect me to insist that my trisection and its proof are valid. And that I so do.
>
> I have never attempted to "Square the Circle." (So delay sending for the fellows in the long white coats.) . . . I am enclosing a picture taken in recent years. I was born May 3, [ ]. I hope to receive communication from you.

The picture showed a blue-eyed, light-skinned, white-haired man, a type common in the Irish and others with Celtic blood. He looked like W. C. Fields, but free of malevolence and with a shrunken nose.

Mr. F. lived in a small midwestern city. His house was south of the center of the city, but near it. Two blocks away was a main thoroughfare, lined with gas stations, fast-food outlets, and the other things that line busy roads, but it was quiet in Mr. F.'s block. It was shaded by old trees and had on it large houses which had the look of being built around 1910. They appeared for the most part to be well kept up, but there were signs of the beginning of the slide into decay. Across the street from Mr. F., a house had been converted into two businesses, a beauty salon and a Peek-a-Boutique. The rest of the block was still residential.

When I rang his bell, Mr. F. came out on the porch to meet me. The house, he said, was in too much of a mess for me to come in. He was short, about five feet four inches, and portly. He was dressed in a gray suit and had on a spotless tie. The suit, though, had spots and did not fit well. Mr. F. later told me he had bought it at a rummage sale for very little—about the retail price for a new necktie. He said that he had to make a telephone call, and before he did he gave me that day's copy of the local newspaper to read. There was no place to sit except the porch railing. He was gone only a short time, so I had no chance to do more than glance at the newspaper.

The call was in fact to the newspaper. Mr. F. had prepared a press release:

Dr. Underwood Dudley, professor of mathematics at DePauw University in Greencastle, Indiana, stopped in [this state] while on a University Program trip that takes him to the Library of Congress in Washington, D.C.

He is shown visiting with [E. F.] of [this city] with whom he shares some mutual interest.

Mr. [F.] designed a mathematical construction seeking to trisect an angle in conformity with a Euclidean limitation that disallows the use of marks or calibrations on a straightedge.

[F.] had heard but disregarded affirmations that the Mathematical Community has proven that absolute Euclidean trisection is impossible.

Dr. Dudley programmed [F.]'s designed construction through the University computer and the results of that programming, according to Dr. Dudley, show that [F.]'s designed construction is an extremely accurate approximate trisection, and Dr. Dudley states that [F.]'s accomplishment is one that not many can perform.

[F.] views these results with mixed emotions. He had hoped for confirmation of his original theoretically evolved construction as a logically perfect mode of trisection. Nevertheless he is very grateful that Dr. Dudley's work on the computer rated his designed construction so very favorably.

He was going to take it to the newspaper office and wanted me along, I suppose, to vouch for the contents of the release. It contained no falsehoods, but I am not sure it expressed Mr. F.'s true feelings. We drove downtown in Mr. F.'s large old automobile, which had defective turn signals and a noisy and smoky engine. After walking upstairs to the newspaper office, Mr. F. presented himself, his release, and me to the city editor. The city editor, who seemed to me to be too young to be a city editor, hemmed and hawed and did not respond to Mr. F.'s very broad hints that maybe a photographer ought to be called in (the second paragraph of the release assumed the existence of a picture). The city editor got out of his difficulties by saying that he would have to talk with another editor who was not there at the moment, but he kept the release.

Mr. F. was prepared for having no picture taken at the newspaper office, because our next stop was at a photographer's studio. That was after Mr. F. had walked two blocks past where he had parked his car because he was so engrossed in conversation—monologue, actually. The photographer was evidently not used to requests like Mr. F.'s, but he agreed to take one black-and-white picture, no proofs, and deliver a five-by-seven print for the rate he charged for taking passport pictures. After having me roll down my sleeves, remove objects from my shirt pocket, and brush my hair, he took two shots. He would have the print ready the day after tomorrow, he said, but Mr. F. insisted on tomorrow, and he got his way.

After leaving the photographer's, we went to where we could talk about his trisection, a hospital. First, we went to the hospital's cafeteria, where Mr. F. ate most of his meals. Judging by its prices, the cafeteria was subsidized, but the hospital authorities must not have minded if some outsiders used it. Mr. F. insisted on buying me a cup of coffee and fetching me a glass of water. He treated me with so much deference as to make me uncomfortable: a man of seventy-eight should not be calling someone quite junior to him "sir," nor should he be holding doors open for him. But courtliness like that was consistent with his letter-writing style. From the cafeteria we went to the lobby of the hospital, where, undisturbed, Mr. F. was able to talk, with a few interjections from me, for almost three hours.

> I grew up on a farm and went to a little one-room schoolhouse. You know, the little country one-room schoolhouse where they alternated fifth and seventh grades one year with sixth and eighth the other. We were just common people, just plain people, but even in grade school I found that some of the students would come up to me with simple little problems, and it amazed me they would come to me asking for help in such simple matters. Later, in high school, I think some of the students relied on me somewhat for help. I remember we had a problem in physics: if you'd set an object between two mirrors with such and such an angle between them, how many images would you have back and forth between the two? They'd come to me for help to solve that.

He graduated from high school during World War I:

> I started to work in a coal office. When the trains came through, full with these draftees or soldiers, there was a wave of enthusiasm or excitement, and I sat there with a dream of the wide expanse of ocean and all that kind of thing. All the enthusiasm and all caused me to go and enlist in the navy. A youth, he sometimes doesn't stop to consider all the sides of a thing.
>
> I had a chance to study radio, which was in its mere infancy at that time. For instance, the detector was just a piece of crystal, a galena crystal, and a cat's whisker (we called it a "cat's whisker"). You had

to adjust to that, and we received by code. I think the most modern
equipment we had at the time was the diode. We went to school, and
we studied electronic principles, the basic math of electricity. That's
what electricity is, principally, math: laws of math, laws of electrical
current, and so on. I went from there to a very high-sounding thing,
the Harvard Naval Radio School, but it was still elementary; we prac-
ticed receiving code and so on. I went up to the Great Lakes from
there, but the war was pretty well along, and although I made a cou-
ple of trips across to Brest to bring soldiers back, why then we were
discharged.

After that I went to this school in the South because my brother had
gone there. It was a very religious-oriented school. They were very
fundamental. I stayed there until I became ill in my last year. Probably
my name was in the yearbook, but I don't know. I would have had a
bachelor of oratory and a bachelor of science degree. It was the strain
of trying to sell books all those summers and all that sort of thing, to
make my way through.

Mr. F. had told me earlier that in summers, teams of college students would be
recruited to sell religious books, door to door, mostly in rural areas, and he had
done this for three summers with varying success. Pennsylvania was very good,
he said, but New York was not: "It was all dairy." Either dairy people were not
readers of religious books or the dairy industry was undergoing a depression
at the time.

Mr. F. had had some ambitions to enter the ministry, and during our talk
he frequently mentioned religious and spiritual matters, but he said that he did
not have any regrets.

I seemed like somehow or other I couldn't visualize going out into the
world with the emotional empowerment to present the sermons the
other fellows did. I just couldn't get into that emotional state, and
it bothered me somewhat. I guess I got depressed. I didn't have so
much the idea of wanting to be in the ministry, but I thought maybe it
was required of me. Maybe it was providential that I didn't go on and
result in a failure in the ministry. Maybe I wasn't cut out for it. I was
considered too rational by some of the students. I didn't present that
emotional appeal; I just couldn't do it and enter into it the way they
did. I wanted to reason things out, probably too much, and of course
when you do that, that kills the emotional side of the presentation.

His school, a small and respectable liberal arts college, is still educating and
graduating students and probably emphasizes reason more now.

So after that then ... well, I have to skip some, make some skips in life when I was ill.

Mr. F. had earlier told me that he had been married, but only briefly. He was married, he said, when he was in the hospital and very ill. When he recovered and left the hospital, his wife left him. It was natural for me to infer that his wife had married him in the hope that he would die and she would collect the proceeds of his life insurance. But Mr. F. did not say that.

I finally wound up reading meters and installing meters, and among my different employments at that time there was a period when I was a troubleshooter for [an electrical utility]. In the course of things I arrived at [a large midwestern city] and worked as a salesman for [another utility]. I worked in a golf club for a while. In [the large city], I was working for [an industrial concern] up there on a grinding machine, and I decided to take a civil-service examination. I got through so fast on the problems on fractions (I did all the work and had some time left over) that they passed me into electronics. I got a job then with [an electrical firm] and was an inspector in radar, and then finally I ended up back here—my folks were getting elderly. I worked as an inspector, doing routine work in electronics and meter testing and so on. That's just a rough outline.

The time goes really fast. Somebody said it seems like you turn around twice and you're old. Life is just a kind of a shadow of eternity, as the Bible says. It just goes so quickly. It goes by so fast. I think it's regrettable that we go through life and we learn so much better how to have lived it, but we can't go back and start over. Knowing what we know now we feel like we could really get somewhere.

Of course, we don't want to look on the dark side all the time, do we? Sometimes we need a little outside help, maybe, to keep us in the right mood. Maybe we should deliberately seek those things that cheer us up and that keep us in a happier vein.

When the subject of the trisection came up, Mr. F. told me how he had gotten started.

Back in 1931 there was an item in the local paper. A professor in an eastern university (he happened to be a priest) presented a solution to the problem of trisecting the angle.

Father C.! It must have been Father C. and his trisection.

So that got my interest. I thought that a method would be found somewhere that would be simple. Somebody where I worked over at the golf club in 1937 in [the large city] told me that some man in London had offered $1,000,000 to anybody who would do it.

This was the first and only time I heard of this, and I have no idea how such a myth could have arisen. In this century, the only prize offered for the solution of a mathematical problem that I know of (except for the small sums Paul Erdős and some people who have followed his example have offered for the solution of specific problems) was the famous Wolfskehl prize for settling Fermat's last theorem. Most of that evaporated in the German inflation after World War I, so since then mathematics has had to be done for its own sake. Nevertheless, word about prizes, real or mythical, gets around. Just a few weeks later I had a letter from a prover of Fermat's last theorem asking for the address of the University of Göttingen so that she could apply for the prize, which, she had heard (correctly), was open to competition until 2007.

Mr. F. agreed with my incredulity about the million dollars:

I don't know why he would do that unless he was terribly involved with the subject and wanted to prove himself right. That might have been the reasoning to do that: to save face or something. I don't know. I know one thing: if a man did come out and develop a trisection it would bring about quite a loss financially and in prestige, probably, to some of the people who for many years had been teaching it's impossible.

I worked on it just once in a while, sometimes with large intervals of time between thoughts about it, but then I got to a place where I began to devote more and more time to it. At one time I thought I had it whipped a few years ago, but then I found a mistake, and I went out in my car and bumped somebody else's rear bumper. [Laughs.] So you see it can kind of depress you if you take it that way. But then I've been elated at times to. Whenever outside influences didn't just wipe the elation away. I think it was the latter part of last year when I finally got the idea that the key to the solution was . . .

Mr. F. gave geometrical details.

Then I proceeded until I hit it and then I started my synthesis.

He circulated his work to 15 or 20 places, with the usual results:

Dr. [—] said he'd study what I gave him. I sent him a revised and corrected copy, and he said that he hadn't had time yet to formulate an appraisal of it. And I've talked by telephone to one or two others, but I just haven't had a firm reply from any of them. Just haven't had any firm reply at all. Yours is the only creditable answer I've had.

I asked whether that discouraged him.

> Well, we should be thankful for what success we have, and none of us probably know the last word in anything. Also, if you ever succeed in anything, you've got to want to, and the degree of success that I have had is due to the fact that I wanted it, very badly. When you want something real badly, you'll exert yourself. We all try to contribute something. I like to contribute something. We'd like to, if we could, produce something which was new. I think that's what justifies a lot of our effort: either improve on what we have or else clear out something else; there's plenty of room, I guess, for improvement.

There was a good deal more miscellaneous and rambling talk. Toward the end of the afternoon, he said

> I can't complain really about the way people are toward me now. I've had some very sad experiences, misunderstandings, that haven't ended up the way that I'd liked. I don't take any comfort in any of the ways I have ... [trailed off]. Even now, for instance—well, I can't go into specifics, but if you have to part ways with somebody, I like to part as friends; I don't like to think it's otherwise. Sometimes you just can't have it that way; that's a depressing thought, but as I say I have no complaint today the way people treat me. People with whom I come in contact today are very nice, and sometimes I hope that something like this will better my relationship or win respect—you know, that you dabble in something and maybe you deserve a little better rating socially than you've been getting.

I left Mr. F. afraid that I had disappointed him—being fairly sure of it, in fact, but not knowing what I could have done to make it otherwise, except to tell him that his trisection was correct, and that would have been as impossible as the trisection. I was glad that he was not offended: next week came a letter with an envelope inside it marked "See Inside!" both on back and on front, so that it could not be missed. Inside was a copy of the picture that had been taken at the photographer's of the two of us, short Mr. F. and brushed-hair me, shaking hands. I heard nothing further for almost a month, but then there arrived a letter with two copies of an article from the local newspaper.

MATHEMATICIAN SCORES NEAR HIT

was the headline over the three-column piece, complete with picture, captioned

> [E. F.], on the right, had a million dollar idea that came within a fraction of working out. He is shown being congratulated by [me], who visited the little mathematician recently.

The "million dollar idea" referred to the nonexistent reward for a solution. The story was accurate and could offend neither Mr. F., me, nor any mathematician. It surprised me that a small-city paper could get a trisection story right, when big-city papers so often get them wrong. The city editor was *not* too young for his job. The story started

> Some people dream of doing the impossible but never get around to trying. A few try, and whether they succeed or not, they have the satisfaction of knowing they have tried.
>
> There is such a man in [this city], a little man 78 years old, whose days are spent alone in his house at [his address] and whose nights are spent alone at the municipal building, where he serves as watchman.

It then summarized his work, ending with

> After all, he dreamed of doing the impossible—and he almost did it.

The letter Mr. F. wrote was in his characteristic style:

> The enclosed News Clippings were not published in the [newspaper] until yesterday, July 21, inst.
>
> They were editorialized to attract general reader interest and yet retain enough technical notation to produce to my satisfaction a good balance of fact & fiction.
>
> As to myself, I have received favorable interest from Others.
>
> As to yourself, I hope that this news publicity and any succeeding or related publicity will widen and add to the good image you must enjoy in the eyes of the Mathematical Community and also generally otherwise. . . .
>
> Hoping that the relative freedom and change from University duties proves both beneficial and enjoyable,
> > I remain,
> > Yours truly,
> > [E. F].

He had gained what he wanted, a little respect. This trisection perhaps had as happy an ending as any can have.

On the surface, there is no obvious trait that all trisectors have in common. The three trisectors show that trisectors can be happy or bitter, they can have a little education or a lot, they can be found in the north or the south, they can be fat or skinny, they can be married or single, they can write well or poorly. However, there is one thing, perhaps not immediately obvious, that trisectors share. What my three trisectors had in common was, I think, ambition: a bright flame burning inside, an urge to be known, a hunger not to be lost among the anonymous millions, a refusal to accept that they, as we, are like ants scurrying around an anthill, a silent yell: "I. I am I! Listen to me!" A. B. painted, wrote novels, did a trisection; C. D. had ruined his professional life trying to get recognition; E. F., though gentle and wistful, nevertheless walked with vigor and purpose and yearned for respect. Most of us can accept that we are ordinary, that the world would not be different had we not been born and will not be different after we die, that even our children (if we have any) will someday die, and that eventually we will be entirely and forever gone from the memory of the human race. But some of us cannot accept that and settle for enjoying the ordinary pleasures of ordinary existence, and must try to have more. Some of us want to leave a mark and make a permanent change. Some of us want to be remembered forever. Some of us want to trisect angles. How foolish we are.

Chapter **4**
A Budget of Trisections

> My intention in publishing is *to enable those who have been puzzled by one or two discoverers to see how they look in a lump.* The only question is, has the selection been fairly made? To this my answer is, that no selection at all has been made. The books are, without exception, those which I have in my own library; and I have taken *all*—I mean all of the kind: Heaven forbid that I should be supposed to have no other books! *(Budget, vol. 1, p. 7)*

I intend to show a lump of trisectors. My intention is broader than De Morgan's: it is not only to enable those who have been puzzled by one or two trisections to see how a lot of them look, but also to discourage trisections and trisectors. I have not only the puzzled layman's interests at heart, but also the busy mathematician's. Were a trisector to look at all of the constructions in this Budget, some perhaps looking very like his, I would hope that at least a faint doubt about the merit of his construction might creep in. The doubt might be great enough to make him stop trisecting altogether, to the good of all concerned: the trisector, his family and friends, the mathematical community, and therefore the world. Even if the doubt only slowed him down a little, it would be helpful. If there were no effect at all, well, there are people who cannot be influenced by evidence and no harm would be done. Human folly will last as long as the human race, but until the race perishes we must always discourage it. If we do not, then the race may perish all the sooner.

> Many an error of thought and learning has fallen before a gradual growth of thoughtful and learned opposition. But such things as the

61

quadrature of the circle, etc., are never put down. And why? Because thought can influence thought, but thought cannot influence
self-conceit: learning can annihilate learning: but learning cannot annihilate ignorance. A sword may cut through an iron bar; and the sev-
ered ends will not reunite: let it go through the air, and the yielding
substance is whole again in a moment. (*Budget*, vol. 2, p. 6)

As De Morgan took all of the paradoxers he had, I have taken all of the
trisectors that I have, with some exceptions. Many trisections are very complicated, or they seem that way. Figures 4.1 and 4.2 are typical, perhaps a bit
more complicated than most. Almost always the constructions can be simplified, making it easier to find out how accurate they are. For all except the simplest, a trigonometrical analysis of a construction quickly becomes too complicated, but it is not difficult to have a computer calculate the locations of
the points in the construction as they are determined and find out how close
it comes to a true trisection. There is one trisection which, when applied to a
60° angle, gives an error of 6′ 20″. Several of the constructions in this Budget
locate the same point, but in ways that are not at all similar. That is part of

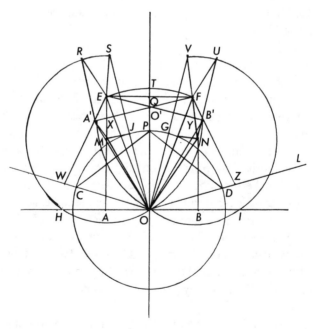

FIGURE 4.1

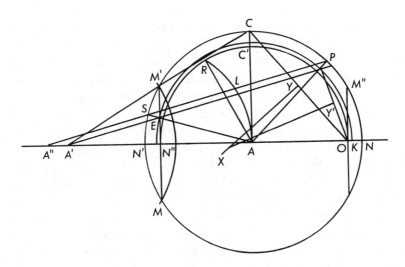

the joy of trisection scholarship: when the computer prints out the 6′ 20″ error at 60°, you know that there it is, the familiar point once again. Fun like this will never replace sex, but we must take our pleasures where we can. However, some constructions are just too complicated to get through, some have instructions which are incomprehensible (or no instructions at all), and some which I thought I had programmed correctly caused the computer to give results which were obviously wrong. These trisections defeated me, and I gave up. After all, no one pays to have trisections analyzed, not even by the world's leading expert. However, there are fewer than ten of these.

There is no harm in not having some trisections in the Budget, because it is in no sense inclusive or exhaustive. New trisections are made all the time, and no doubt someone is at this moment working on one. I do not have all of the old ones either, since trisectors are not required to send their work to me and since trisections sent elsewhere are usually not kept. The trisections of one generation are thrown away by the next, if they even survive that long. However, I think that the genre remains fairly constant over time: the lost trisections of the 1920s are probably similar to those of the 1990s, and the following collection is probably representative of all trisections, past, present, and future. The more than 100 trisections in the Budget are enough: more examples would not yield any more enlightenment.

I think that it is amazing that there are so many different and relatively simple constructions with straightedge and compasses that give good approximations to the trisection. Some of them are astonishingly accurate, and all of them are good enough to fool the eye or the protractor, at least for angles of reasonably small size. Of course this is no surprise, because a trisection that was obviously inaccurate would not be sent out by the trisector; he would go back to his drawing board until he had found something better.

The very multiplicity of constructions will serve, I hope, as a strong indication to a trisector that *his* new construction cannot be correct. "Trisector," they say to him, "wouldn't you agree that if there were a simple way to trisect an angle, there would be only one way? After all, there is only one way to bisect an angle. And don't you think that if there were a way, someone would have found it by now? Look at all of these failed attempts; what makes you think that yours is any different? Are you really that much smarter, or luckier, than all of those people?" We know, and the trisector should know, that it is impossible ever to find a trisection, but to someone who would not or could not grasp what *proof* and *impossible* mean in mathematics, the implied argument may be plausible enough to make him stop trisecting. Unfortunately, many trisectors have egos big enough to let them say, at least to themselves, "Yes, I *am* smarter, or luckier, than anyone else," and no argument will have any effect. The only way to stop such trisectors would be to make trisection a federal crime with heavy penalties and to hire a large group of FBT operatives (the Federal Bureau of Trisections, that is, not yet in existence, but for whose Director I have a nomination, assuming the salary is sufficiently large) to hunt down suspected trisectors. That is not going to happen, so this book is the feeble next best thing.

Almost all of the trisectors in this Budget are genuine trisectors. That is, almost all were convinced that they had discovered a completely accurate trisection. I have not included the attempts of those who were obviously young students, not clear on the impossibility of the trisection. They write things like "My construction looks as if it trisects angles, could you take a look at it, please?" They are not genuine trisectors, and, if they are lucky, they never will be. There are a few nongenuine trisectors in the Budget, but they have other points of interest.

Although many trisectors are eager to spread their work before the world, printing up copyrighted pamphlets and even books and distributing them widely, others never go beyond correspondence. Rather than distinguish between the two classes, I have referred to all trisectors, publicity seekers or not, alive or dead, by their initials only. Most of those who are still living and have trisected only in correspondence have had their initials altered. The dates of all the constructions have been given, but where they were done has for the most part been suppressed. Here is a summary of the locations of the trisectors in this Budget:

Number	Place
11	Place unknown
21	California
8	New York
6	Indiana
4	Michigan and Pennsylvania
3	Illinois, New Jersey, and Ohio
2	Arizona, Iowa, Kansas, Massachusetts, Missouri, and Washington
1	Arkansas, District of Columbia, Florida, Hawaii, Louisiana, Montana, Nebraska, Oregon, Texas, and Wisconsin
3	India
2	England and New Zealand
1	Australia, Austria, Canada, Finland, France, Germany, Greece, Guyana, South Africa, Taiwan, and Turkey

It is no surprise to find California leading the way among the states. The large number of trisections from Indiana and the Midwest is explained by my location in Indiana; the power to attract trisections evidently declines with distance, though clearly not in proportion to the reciprocal of the square of the distance between me to the trisection. The exact law remains to be determined, and I suspect that it never will be found. The world is full of mysteries which will never be explained. In any event, the identity or location of individual trisectors is not important; they are only examples and representatives of a class.

In what follows, AB denotes the line segment joining points A and B and $|AB|$ denotes its length. The trisection point in all constructions is labeled T.

A Pythagoreanism

By B. L. A., 1970

No proof of the correctness of this trisection is offered, only the details of the construction. In Figure 4.3, as in almost all the other figures, angle BOA, the angle to be trisected, is a 60° angle. Draw a circle with center B, tangent to OA, and find C by drawing one-half of angle BOA with vertex at A. Draw horizontal lines at B and D and a tangent to the circle at C. EF is perpendicular to CE and arc FT has center O and radius $|OF|$; its intersection with the horizontal line through B determines the trisection point T.

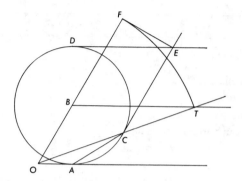

FIGURE 4.3

The trisection would be accurate if only $\sin(\theta/3)$ equaled

$$\frac{\sin\theta}{\sqrt{2\cos^2\theta + 4\cos\theta + 3}}$$

but it does not, and the maximum error for trisecting acute angles is 20′.

Untitled Trisection

By R. R. B., 1983

B. sent his trisection of the 60° angle to the head of the Pure Mathematics Department at the University of Moscow and to the executive secretary of the American Society of Civil Engineers, of which he is a fellow. He did not say whether he had any replies.

The construction applies only to the 60° angle BOA (see Figure 4.4). C, D, E, and F are found by stepping off the radius, $|OA|$, of the circle around the circumference. The arc DT has center G and radius $|GD|$. The approximation, which amounts to

$$\tan 20° = \frac{\sqrt{7} - 2}{\sqrt{3}}$$

is not especially good, the error being 27′.

The mistake in the proof is the common one of trying to make a line do too much: "Join J to C and let it cut 150° arc at J'." If lines could speak, segment JC would cry out, "I can't do it! I can't and still be straight!"

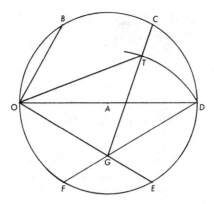

FIGURE 4.4

Trisecting a 60-Degree Arc

By N. L. B., 1927

This is a short note in the June 1927 issue of *Industrial Arts Magazine,* in between instructions on how to build a garden house in a school shop and how to use a sled to find the coefficient of friction. B. applies his method only to a 60° angle, but it can be used on any angle.

In Figure 4.5, C is the bisector of OA, CD is parallel to the bisector of angle BOA, and ET is drawn parallel to OB. The error in this construction is large, being as much as 2° around 36° and 5° around 81°. But it gets small near the angle the author trisected: 30′ at 60°, only 4′ at 63°. Somewhere around 62° it is zero.

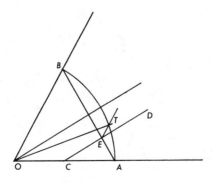

FIGURE 4.5

To: Tri-sect an Acute Angle—Using but Straightedge & Compass

By W. C., 1975

C. described his construction entirely in words and added

Unique, perhaps, but for once I am omitting the (usually) sloppy drafting!

It is in fact unique: no other trisector in this Budget failed to include a diagram. C. was wrong about the drafting because most trisectors take some care with their pictures and some are so beautiful as to qualify as works of art. I have hanging on my office wall a multi-colored hand-lettered sheet drawn with such marvelous care that it resembles a medieval illuminated manuscript.

The construction is more complicated than most. In Figure 4.6, strike off arc CD and make arc EF have a radius twice as large, both with center at O. OC' bisects angle BOA and H is on the chord EF. FJ is perpendicular to OA, and J is on the perpendicular, $|GJ| = |HI|$. Arc KL, center at O, has radius JD. T is on arc EF, $|KL|$ units away from F.

The construction illustrates the danger of working only with angles near $60°$. It amounts to claiming that

$$\sin\left(\frac{\theta}{6}\right) = \frac{1}{2}\sqrt{1 + 4(\tan(\theta/2) + \cos(\theta/2) - 1)^2}\,\sin(\theta/2)$$

and it is very inaccurate except for angles near $60°$. The error at $60°$ is only $12''$, though.

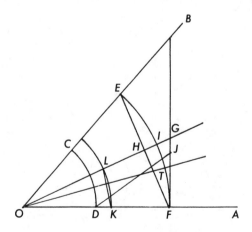

FIGURE 4.6

A Vital Question in Mathematics; Mathematics Out of Joint

By L. S. C., 1892

C. was presented with evidence that his construction was not accurate:

Prof. Harry Phelps, United States Naval Academy; Prof. Thomas S. Barrett, London, England, and other geometers show that Trigonometric calculations fail to agree with above Trisection of Angles.

So there is a choice: either the trisection is wrong or trigonometry is wrong. C. fully appreciated that, so he concluded that he was right and trigonometry was wrong.

Present Text Books of Mathematics are fallacious and new Tables and Text Books are required.

In Figure 4.7, the semicircle has center A and the radius is $|OA|$. DE is parallel to OC, and angle DAC is constructed to be $60°$. F is the intersection of DE and OB: in this figure, angle BOA is $60°$ so F and B coincide; for other angles, they would be different points. C is found by making $|DG|$ equal to $|EB|$, and T is the intersection of the semicircle with GA. C. restricted the angle BOA to values between $60°$ and $90°$; in that range, the construction is accurate to $27'$. It can be applied to smaller angles, but it is not very accurate (the error is more than $1°$ for a $24°$ angle), and the error increases rapidly as the angle increases beyond $90°$.

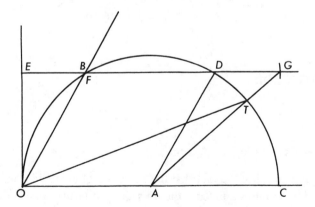

FIGURE 4.7

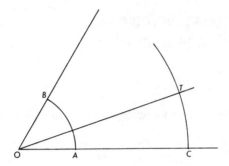

FIGURE 4.8

Trisection

By H. A. C., 1931

On reading about Father C.'s trisection, C. sat down, did his own, and included it in a letter to the editor of the *Indianapolis Star,* from which it was clipped by a then member of the mathematics department at my school and put into a folder, where it yellowly remains today.

The construction is very simple (Figure 4.8): draw arc AB, make $|OC| = 3|OA|$, draw an arc with center O and radius $|OC|$, and draw another with center C and radius $|AB|$ to locate T. It is also very inaccurate, with errors exceeding 1° for acute angles.

Trigonometrically, it says that

$$\tan\left(\frac{\theta}{3}\right) = \frac{2\sin(\theta/2)\sqrt{9 - \sin^2(\theta/2)}}{3 - 2\sin^2(\theta/2)}.$$

A Construction

By N. B. C., 1982

Here is a leap of logic:

I am fully aware of the fact that the trisection of an angle is impossible and that Galois and Co.'s theories and proofs are conclusive and 100% correct in their own right. I couldn't agree more.

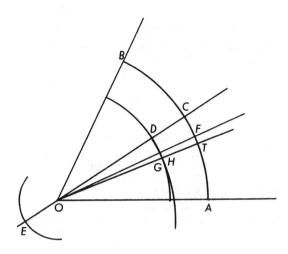

FIGURE 4.9

BUT my construction is *NOT* a trisection in that traditional sense. *It is the expansion of angles in the ratio* 3 : 4, resulting in the expansion of 4/4 of an angle to 4/3 of its original size. Then $(4\theta/3) \div 4$ cannot be anything but $\theta/3$! (Precisely too.)

C., who was a teacher of mathematics, did not see the inconsistency.

In Figure 4.9, strike off arcs with center O and radii 1, 3, and 4. The bisector OC determines C and E. Make angle COF one-eighth of angle BOA by two more bisections, and find G so that $|DG| = |CF|$. H, which is very close to G, is the intersection of EG with a circle, center E and radius 4. C. extends OH to get T. The maximum error for acute angles is always less than 47″, except for some strange large errors for angles less than 7°. Even for angles up to 180° the error is less than 6′, so this is quite an accurate trisection.

Although we continued to correspond for a time, C. remained convinced of the correctness of the construction.

A Solution Making Possible the Trisection of an Angle

By E. C., approximately 1959

This is the only example I have of a trisector who used calculus in his trisection, even unto partial derivatives. Like the previous author, he is a counterexample to the theorem that knowledge of higher mathematics provides immunity to the trisection disease. He claimed to have found a relation between

$\sin\theta$ and $\sin(\theta/3)$ which uses only square roots; if such a thing were possible, then the trisection could be accomplished with straightedge and compass alone. Here is the relation: if $s = \sin\theta$ let

$$z = \frac{3 + 3s^2 - 9s - \sqrt{9 - 6s^2 - 3s^4}}{12s}$$

and let

$$A = \frac{4s^2\sqrt{9 - 6s^2 - 3s^4}}{3 + s^4 - (1 - s^2)\sqrt{9 - 6s^2 - 3s^4}}$$

then $\sin(\theta/3)$ is, he said, the quotient of

$$\left(9 + 9s^2 - 2Asz - \sqrt{9 + 9s^2 - 2Asz}\right) - 4s(9s - Az)\sqrt{6 + 3s^2 - Asz}$$

and

$$4(9s - Az).$$

The derivation is complicated, and I did not try to follow it. It is helpful to have a trisector with some mathematical training, as this one: because of it, he is incapable of the utter incoherence that other trisectors sometimes achieve. This can be taken as an argument for mathematical training. Unfortunately, it is not conclusive, since it is impossible to refute the counterargument that a mind able to comprehend third-semester calculus and manipulate complicated expressions is likely to be incapable of incoherence, mathematical training or not.

Though his derivation was hard to follow, his claim was clear and hence easy to check. Since the check showed that the claim was wrong, there was no need to go through the derivation to find the error. The approximation is very good, with a maximum error for acute angles of no more than 3', which occurs near 45°.

It would not be easy to translate the preceding equations into a straightedge and compass construction, so there is no diagram for this attempted trisection.

Trisection

By C., date unknown

No one's files are perfect, and mine are no exception. On the back of a sheet containing the results of my school's 1983 faculty elections appears the trisection of Figure 4.10. Draw arc CD and bisect it to get E. Draw a semicircle

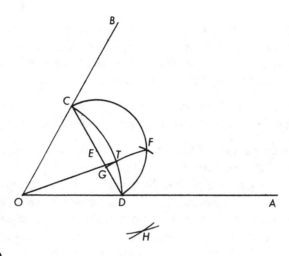

FIGURE 4.10

with CD as diameter and find F by drawing an arc with center D and radius $|DE|$. Trisect CD to get G and locate H at a distance $(4/3)|CD|$ from both F and G. Then T lies on the arc CD, where it is intersected by an arc with center H and radius $|HG|$. I have not yet gotten around to running this construction through the computer, but my trisection intuition tells me that it is probably pretty good. I would bet that the maximum error is less than $8'$ for acute angles.

Trisecting Angles Geometrically

By D. C., 1981

This is a 16-page pamphlet with opposite pages printed in English and Turkish. C. had heard of Wantzel's proof of impossibility, but he wrote

If certain conditions are eliminated, the proof becomes invalid. And that is what happened.

He gave no details. He also had read, or looked at, a text on ring theory that contained an impossibility proof, and so, just as with L. S. C., five authors ago, 100 years earlier, and half a world away, the conclusion is that mathematics has to give way to the trisection:

Thus, it becomes a necessity to revise ring theory.

Trigonometry, ring theory: no part of mathematics is safe from a trisector.

I did not try to master the details of his complicated trisection, because he gave a formula for $\cos(\theta/3)$ in terms of trigonometric functions of $\theta/16$ (also complicated), which cannot be correct. When I sent him a table of the size of the error in his construction, he answered that his proof could be refuted only

> on the basis of the elementary geometry principles, and not by the numerical evaluation of the trigonometric expressions related with it.

This is a common response. Many trisectors say that since they did not use trigonometry to make their construction, you may not use trigonometry to examine it. Some assert, quite explicitly, that since their constructions use straightedge and compasses only, they can be refuted only with straightedge and compass. Similarly, circle-squarers, when told that their construction gives $\pi = 3.125$ or whatever, will sometimes reply by saying that their construction had no numbers in it. They do not realize that truth remains truth whether you use it or not.

Trisection

By R. C., 1964

I said that almost all the trisectors in this Budget were genuine trisectors, but it would not be complete without a sample from a nongenuine one, and here it is:

> I am in the eighth grade and very interested in geometry. I was told that it was impossible to trisect an angle using only euclidean tools. I think I have found a way to do it. So far my teachers and their colleagues have not been able to prove or disprove my theory, so I am writing to ask if you could help me. Enclosed is an example of the method I use.
>
> Thank you,
> [R. C.]

Definitely an ingenuous trisector rather than a genuine one.

In Figure 4.11, OC is the bisector, D is the midpoint of OA, and T is the midpoint of DC. It is a simple construction giving the

$$\tan\left(\frac{\theta}{3}\right) = \frac{2\sin(\theta/2)}{1 + 2\cos(\theta/2)}$$

point found by others.

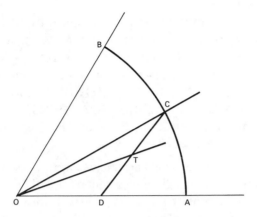

FIGURE 4.11

It is rather sad that neither C.'s teachers nor their colleagues could do the rather elementary trigonometry needed to show that, but trigonometry is not in the eighth grade curriculum and what you do not teach you tend to forget. One hopes that C. had better luck in high school.

File

By S. I. C., 1968–75

This trisector has a file folder all to himself because he wrote voluminously to me, to famous mathematicians, to teachers of mathematics, to anyone who would write back and many who would not. He trisected only the 60° and 120° angles and his diagrams are very complicated, so I will omit his construction.

When I asked him about his trisection history, his reply included

My friend, it would take a book to tell about my involvement with this. I DID NOT DECIDE to get into the work of seeking the solutions, I WAS PUSHED IN. It started back in 1968. My attention was called to a, "Ancient Mysteries", because of some doodles I had made. Somehow I knew HOW to relate them to writings THAT WERE SHOWN TO ME AFTERWARD, and one book stated that, "The relation of the diameter of the circle, etc., was the, "KEY", to the Mysteries. So off I went to find that, "KEY".

Was my efforts worthwhile? Many friends offered me exceptional opportunities, if I would consider returning to the business world. My answer. "NOT IF I COULD MAKE 10, million DOLLARS." Likewise I DID NOT EXPECT TO GAIN ANYTHING, I HAVE ALREADY RECEIVED MORE THAN ANYONE COULD ASK FOR. As to what I expect now, it isn't what I expect but what I hope and pray for, THE OPPORTUNITY TO SHOW A FEW OF THE SECRETS I HAVE DISCOVERED, SO OTHERS MAY BENEFIT BY WHAT WILL BE SHOWN. Based on what I have already said you KNOW I would do it all over again.

Knowing the society we live in I am trying to use the, "SOLUTIONS" as a door opener. If I am given recognition for SOLUTIONS, it will afford me the opportunity TO SHOW OTHER, "SECRETS", which is really what the unsolved problems is all about.

The Trisection of the Angle; The Trigonometric Function of One-Third of an Angle in Terms of the Functions of the Angle; The Insertion of Two Geometric Means Between a Line and Another Twice as Long; The Duplication of the Cube

By J. J. C., 1931

Father C. has been mentioned elsewhere. In this 29-page pamphlet, written in the style of textbooks of Euclidean geometry, not only is the angle trisected, but the cube is duplicated.

Figure 4.12 gives a simplified version of the diagram which Father C. used to demonstrate his trisection; after I drew it, I realized that the construction could be done even more simply (Figure 4.13). Draw a line through B parallel to OA; mark off the distances $|AC|$ and $|CD|$, each equal to $|OA|$; draw the arc DE with center C and radius $|CD|$; drop a perpendicular to OD from E; and draw the arc FT with center O and radius $|OF|$ to get the trisection point T. It is an exercise in elementary trigonometry to show that the construction is equivalent to asserting that

$$\sin\left(\frac{\theta}{3}\right) = \frac{\sin\theta}{2 + \cos\theta},$$

an approximation that has been found by others in this Budget.

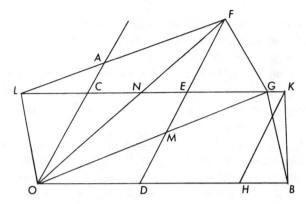

FIGURE 4.12

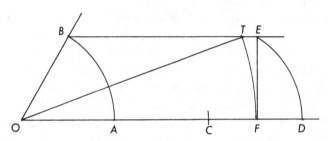

FIGURE 4.13

Encompassment Geometry (1968)
How to Trisect the Angle—Using Only a Compass and a Straightedge, in the Annular Newsletter (1972)

By S. W. C.

Here is an example of crank calling out to crank. The *Annular Newsletter* (later published as *Stonehenge Viewpoint*) had as its primary purpose promoting the theories of I. N. Vail, the main one being that in prehistoric times the earth was surrounded by a halo of ice crystals. The halo produced atmospheric effects that were recorded, Vail maintained, in monuments like Stonehenge, in prehistoric art, and in many other places. The *Annular Newsletter* devoted large amounts of space to diagrams showing how the halo produced icebows (like rainbows) which were memorialized by prehistoric people. Vail's canopy theory may or may not be correct (most likely not), and it is hard to understand why anyone would devote his energies to promoting it since it seems quite irrelevant. If the whole world tomorrow embraced the theories of Vail, science, technology, and life would continue much as before, and only the editor of *Stonehenge Viewpoint,* who, by the way, wrote very well (cranks are not always uneducated), would benefit.

C. may have been a Vailian, since to be a crank in one way much increases the chance of being a crank in another way, but he may only have been looking for an outlet for his trisection and was lucky enough to find one. In the issue of *Annular Newsletter* after the one containing the trisection there were three letters refuting it. *Encompassment Geometry* then disappeared from the list of books available from Annular Publications, and I have seen no mathematics in the pages of the journal since then.

The construction (Figure 4.14) depends on locating a point G which can then be used to trisect any angle. To find it, construct a 45° angle COA and a 15° angle DOA and erect a perpendicular to OA at A. Draw CE parallel to OA, trisect the segment EA to get F, and extend DF to get G. Now, given an angle BOA to trisect, draw BH parallel to OA, trisect HA to find I, and extend GI to the trisection point T. The construction is exact for a 45° angle, but for a 60° angle, it is off by 17′ and the error increases rapidly for larger angles.

C. had heard that Wantzel had proved the trisection impossible, and this is how he got around that:

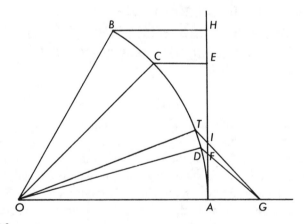

FIGURE 4.14

However, in 1931, Kurt Gödel published his "Incompleteness Theorem," later known as "Gödel's Proof" which shows mathematics to be non-ending and is in fact, an infinite regress of richer and richer systems.

One of the implications of Gödel's great work must then be that all negative proofs of impossibility are limited to the extent that a process is impossible only within the boundary of presently known mathematics.

So in a world of infinitely expanding mathematics there always remains the possibility that the invention of a new mathematical process will render the so-called impossible, possible.

A nice try, but not quite successful. Gödel proved that in any axiomatic system, such as plane Euclidean geometry, there will always be true statements that can never be proved and false statements that cannot be disproved. If the system is enlarged, it may be possible to prove something that could not be proved before, but there will still be things which are undecidable. No matter how far you go along the infinite regress (progress, actually) of richer systems, there will always be nonprovable truth. But once something is proved it remains proved in any larger system, so Wantzel's proof will never be overturned by any new mathematical process.

Trisecting the Angle

By H. C., in *Mechanix Illustrated*, 1966

Mechanix Illustrated was noncommittal on whether the trisection had actually been accomplished:

A reader tells us (at length) that he has trisected the angle—something not done in 3,500 years of trying. He challenges *MI* readers to find the fallacy in his trisection. So here it is (see below). Address your replies to Trisection c/o *MI*, etc. We will hold all letters for Mr. [H. C.], who is something of a mystery man but, we swear, has actually been seen by us. We are assured that the skeletal construction and proof which follow can be followed to the end.

The bad effect of this editorial misjudgment is that people will vaguely remember that something about the trisection appeared in something that they read, and they may have the impression that it is still an open question and

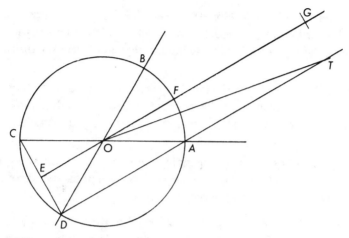

FIGURE 4.15

one that might repay their time spent on it. *Mechanix Illustrated* was a little irresponsible.

The construction, considerably simplified, is in Figure 4.15. EF is the bisector of the angle BOA, and $|FG|$ is the same as $|EF|$. The trisection point is located by extending DA and finding its intersection with an arc with center E and radius $|EC|$. It is quite close for all angles from 0° to 180°, with maximum error 4', and for angles up to 69° the error is less than 1'. A simple construction; if such things were valuable, this would be one of the more valuable.

The Impossible Solved

By E. S. D., 1975

D., an accountant, wrote the publishing firm of Springer-Verlag, "Attention: Mr. Verlag." He probably assumed that Springer was long dead. He was very concerned with copyright, every page being marked with "©by [D.] Enterprises Ltd. World Rights Reserved" or something similar, and one page had, as its sole content,

ANY PERSON USING THE TRI-SECTION INFORMATION; AS TO THE DRAWINGS, WORDS, METHODS ETC.; FOR PRIVATE USE, ZEROXING FOR SCHOOL USE, OR ANY OTHER USE WITHOUT THE WRITTEN CONSENT OF THE AUTHOR OR PUBLISHERS OF

THIS BOOK WILL BE SEVERELY DEALT WITH IN THE COURTS. THE
AUTHOR'S NAME AND ADDRESS IS; . . .

On the whole, it seems wisest not to include any details of the trisection. I
strive to stay out of the courts as much as possible. D. has also duplicated the
cube, squared the circle, and—something unusual—proved "that a circle is a
12000 sided figure."

Trisection of Arcs and Angles with Compass and Straight Edge Only

By G. W. D., 1975

The author of this nicely printed 32-page pamphlet is the only trisector in
this Budget whose construction came from without and not from within:

> The facts I publish here have been revealed by the Great Mathemati-
> cian, for I could get no help from any book; neither would scholars
> aid me because of built in convictions.

In Figure 4.16, AC is drawn perpendicular to AO and D is located equidis-
tant from O and C. The circle has center D and radius $|DO|$. The line through
A parallel to OB intersects the circle at T. The maximum error is about $9'$ for
angles less than $90°$. Divine revelation seems to be no better than any other
method for producing accurate trisections.

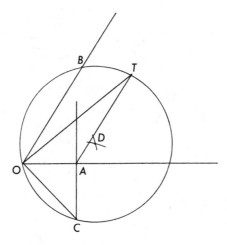

FIGURE 4.16

Untitled Trisection

By N. R. D., 1975

This trisector, started on his career by a recollection of a geometry class, gave some of the history of his construction:

At first, I was really ignorant. I tried to use segment BC across angle BAC and trisect it. Then I tried to adapt Archimedes' method so you wouldn't have to put marks on the straightedge. Another failure.

Then I came upon ...

the following construction.

In Figure 4.17, C and D are on OB and OA extended at a distance $|AB|$ from B and A, respectively. E is on an arc with center B and radius $|AB|$, $(1/2)|AB|$ from C, and F is on the other side of the arc on EB extended. G and H are similarly determined. J is on arc DB, $(1/2)|AB|$ from G; K is on arc CA, $(1/2)|AB|$ from E; and L is on the same arc, $(1/2)|AB|$ from K. Finally, T is the intersection of FL and HL. For all that work, the construction is not especially accurate, except near $45°$.

As often happens, the author sent his construction off to many places, and he got many replies.

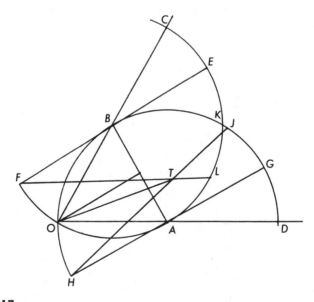

FIGURE 4.17

Unfortunately, my construction was neither proved nor disproved by any of the other universities, although I did learn some very interesting facts about attempted trisections.

It was not proved because it could not be proved, and the author did not think it was disproved because he could not understand the disproofs. What a waste of time!

Angle Trisections

By J. D., 1974

D. submitted a manuscript to the *American Scientist*.

It has been proved impossible to trisect a 60 degree angle with a straight-edge and compass alone. However the following construction will prove that the 60 degree angle can be trisected while limited to only a straightedge and a compass.

The construction consists of taking an angle and tripling it. The original angle is then one-third of the tripled angle. That the author should think that was worthy of note, much less six pages of manuscript, is an illustration of the combination of ignorance and arrogance that many trisectors have. If the original angle was 20° then the 60° angle has been trisected. In other words, to construct a 20° angle, all you need is a 20° angle to start with.

Granted the author knows not what radius will produce an angle which measures 60 degrees, but there is such a radius and one may by chance trisect a 60 degree angle.

However, the chance is zero.

Untitled Printed Sheet

By H. De M., Michigan, 1931

The author's name was *not* De Morgan. In Figure 4.18, arcs DE and GF can be struck off with any convenient radius. Let OC be the bisector of angle BOA and draw lines through E and D parallel to OC. J is the intersection of the bisector of angle HOA and the arc GF. T is the intersection of JK and the segment through D parallel to OC.

If done accurately, the construction is quite good—astonishingly so. If the distance $|OE|$ is about $(4/10)|OB|$ and $|BG| = |OB|$, the maximum error for

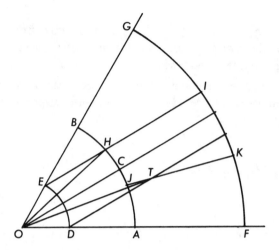

FIGURE 4.18

angles less than $130°$ is a mere $20''$. By varying the radii of OE and OG, the accuracy can be improved. For example, if $|OE| = (0.36)|OB|$ and $|BG| = (1.1)|OB|$, the construction has error less than $1''$ for angles between $0°$ and $135°$. This is my all-time champion construction, though the accuracy would be very hard to get in a drawing, since T is the intersection of two very nearly parallel lines. Even with a very sharp pencil, such a point is hard to determine.

D. wrote to a well-known geometer, who pointed out the error in the proof in a paragraph. D. replied

You missed the point.

Why don't you visit me and let us both have a pleasant dissertation on the Trisection!

The well-known geometer commented to me

They always say this. Obviously I beat a hasty retreat at this point.

It was the best thing to do for all concerned.

Trisection of an Angle

By R. A. D., 1969, 1974

D. was inspired to work on the trisection by reading about Father C.'s trisection in 1931. Thus does the foolishness of one generation infect the next. Freedom of the press should stop short of allowing newspapers to print articles on trisections; it is much the same as throwing cholera bacteria into reservoirs, a thing not generally allowed. The trisection virus lay dormant for decades, but flared up in the 1960s when D. found his construction. Since then he has corresponded with many mathematicians.

Figure 4.19 is a copy of D.'s own diagram, and it is typical with its clutter of lines and circles. I will not describe how the points are determined because the construction can be simplified, as is almost always the case. In Figure 4.20, C is the foot of the perpendicular from B to OA. $|OD| = |OE| = |OA|$, F is the midpoint of DC, and G is the midpoint of EC. H is the intersection of arc AB with an arc with center G and radius $|GC|$. I is on OH, with $|OI| = |OC|$. J is the intersection of a line through I parallel to OA with an arc with center G and radius $|GC|$. It is only a coincidence that in the figure the point J lies on

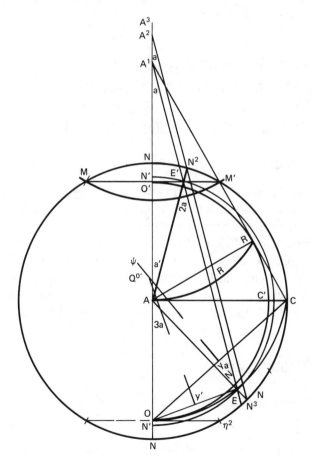

FIGURE 4.19

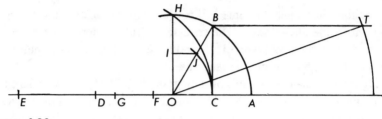

FIGURE **4.20**

OB. Finally, T is on a line through B parallel to OA and an arc with center O and radius $2|OA| + |OJ|$.

For angles between $0°$ and $60°$ the error does not exceed $46''$, but it increases rapidly thereafter, reaching $30'$ at $105°$.

The Geometrical Problem Solved; A Manual for Scientists and Students; How to Trisect or Divide an Angle into Any Number of Equal Parts or Fractions of Parts

By W. D. D., 1892

D. explained his construction in a 16-page pamphlet but did not attempt to give a proof of its correctness. He thought that the trisection was important, and he did not understand that it was impossible.

The construction (Figure 4.21) is simple: construct three concentric arcs with $|OC| = |CE| = |EA|$. Draw DA and FE; their intersection, along with C, gives the trisection point T. Along with simplicity goes inaccuracy: the error increases steadily as the size of the angle to be trisected increases, and it is more than a degree at $90°$. The trisection art has advanced since 1892.

How to Trisect Angles

By U. D., Greencastle, Indiana, 1977

D. wrote

The ancient Greeks, what did they know? You want to be treated by an ancient Greek doctor? You going to ask an ancient Greek as-

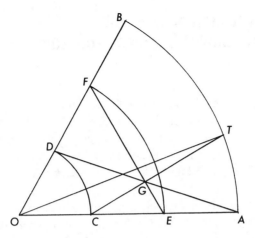

FIGURE 4.21

tronomer about black holes? The ancient Greeks thought everything was made out of earth, fire, air, and water, that's how good they were at chemistry. Primitive! They were primitive! We know more now. Now we can split atoms, go to the moon, look inside people. Now we can trisect angles. Like this:

In Figure 4.22 OC is the bisector, OD the bisector of the bisector, and the chord DA is trisected at E. Angle TOE is equal to angle DOA. The maximum error is 6′ for acute angles. Rather primitive, not to mention crude.

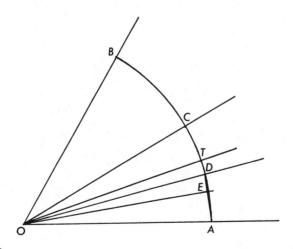

FIGURE 4.22

The Thrice Division of an Angle by the Simple Use of Compass and Ruler

By J. E., 1973

This is the trisector who spent 12,000 hours on the trisection.

To get the construction in Figure 4.23, draw OC, the bisector of angle BOA. Divide OB and OA into nine equal parts, thus determining D, E, F, and I. Draw a line through D parallel to OB to get G and extend OG to get H. Draw IJ parallel to AB and find T by drawing an arc with center I and radius $|JC|$.

As might be expected of such a complicated construction, it is fairly accurate, with a maximum error of 8′ for angles from 0° all the way to 145°. I suspect that the author spent a good part of his 12,000 hours finding H and thinking it was the trisection point; the rest must have gone in the revision from H to T.

Since the author spent so much time on his construction, I put considerably more effort than usual into finding what it amounted to trigonometrically. It is this: $\tan(\theta/3)$ is the quotient of

$$(\sin\theta)\left(1608 - 676\cos\theta + 68\cos^2\theta - \sqrt{D}\right)$$

and

$$(6 - \cos\theta)\left(168 + 100\cos\theta - 68\cos^2\theta + \sqrt{D}\right)$$

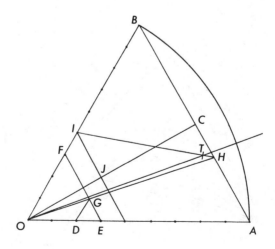

FIGURE 4.23

where

$$D = 52200 + 27600 \cos \theta - 20750 \cos^2 \theta + 3650 \cos^3 \theta - 200 \cos^4 \theta.$$

The General Trisection of an Angle

By N. R. E., 1956

E. knew about trigonometry, the history of the problem, and Wantzel's result. Nevertheless,

> Not being satisfied with [demonstrations of impossibility, I] began the present work, and after nearly three years on it . . .

In his proof that his trisection is exact, E. used Desargues' theorem, a result which most college mathematics majors have never heard of. The theorem is so pretty that I cannot resist stating it here. If you have, as in Figure 4.24, two triangles such that lines joining corresponding vertices meet at a point, P in the figure, then the intersections of the extensions of corresponding sides will meet in three points which lie on the same straight line. How in the world did Desargues ever find that out? You can see the appeal of the theorem to the trisector: three pairs of vertices, three pairs of lines, triads all over the place. But the trisector who is familiar with projective geometry is rare, and this is surely the only time a trisector has cited the theorem. E. is another counterexample to

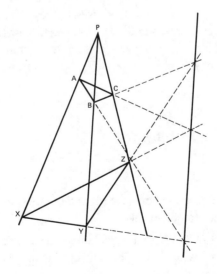

FIGURE 4.24

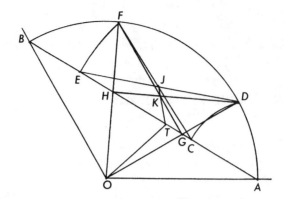

FIGURE 4.25

the theorem that if you know mathematics beyond the most elementary level, you will not trisect.

If one knows mathematics, one is able to use it, and E.'s construction is unusually accurate, with a maximum error of only 3′ for any angle from 0° to 230°. It would be hard to carry out, though, since as E. notes, "the edge (JK) is generally small." It is nevertheless remarkable and provides evidence that the better a mathematician a person is, the better a trisector he will be. As teachers of mathematics have always told their students, mathematical training is good for you no matter what you are going to do.

The edge JK is so small that in Figure 4.25 the angle BOA to be trisected is 120° rather than 60° so that it will be visible. Arcs CD and EF are struck off with center A and convenient radii. The intersection of OD and BA gives G; the intersection of OF and BA gives H; the intersection of ED and CF gives J; the intersection of FG and DH gives K; and the intersection of JK and AB gives T.

Tri-section of an Angle

By B. A. F., 1974

F. had not mastered the language of mathematics: in his proof of correctness there appeared statements like

By the swing of DB to D, with a gradual increase in the rate of cutting arc ID, it will arrive at Z with a built up cutting rate for the last degree which is here called X.

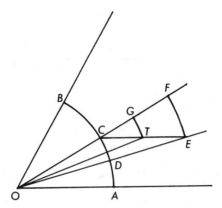

FIGURE 4.26

It is not in the style of Euclid.

In Figure 4.26, OC is the bisector of angle BOA, and OD is the bisector of angle COA. Draw a line through C parallel to OA and let E be its intersection with OD extended. An arc with center O and radius $|OE|$ determines F. G is the midpoint of CF, and T is found by drawing an arc with center O and radius $|OG|$. The construction is accurate enough so that the error would not be noticeable on an ordinary drawing: it increases with the angle to near $14'$ at $90°$.

Trisection

By W. S. F., in the Cincinnati *Post and Times-Star*, 1966

This newspaper story starts

Trisecting an angle using only a straightedge and a compass has caused geometry students many a sleepless night ever since it stumped mathematicians in Plato's time.

Fine so far, but it goes on

Solutions published from time to time all have been disproved, and many mathematicians say it just can't be done.

Wrong, *Post and Times-Star*; not many, *all.* Newspapers seem to be as unable as trisectors to get it through their heads that the trisection is impossible and is not a matter of opinion. Clearly, journalism majors should all be required

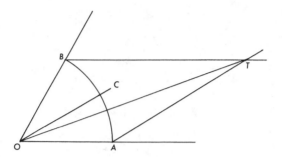

FIGURE 4.27

to take several courses in mathematics. The story, complete with diagram and four columns wide, told how W. S. F., a real estate developer,

> got interested in the problem as a young man. . . . He's been working on a solution for the last few years. . . . He's copyrighted it and plans to submit it to mathematics organizations and scholarly journals. So far, he's found only tentative acceptance.

I hope the story caused some kind mathematician to persuade F. not to make a further fool of himself. His picture showed a well-dressed man in his fifties. It is not usual to have so young and prosperous a man do the trisection.

His construction is not very good. That is not surprising: he had not spent enough time on it (developing real estate keeps you busy) to do better.

In Figure 4.27, OC bisects angle BOA, and T is the intersection of a line through A parallel to the bisector and a line through B parallel to OA. It is, once again, the construction that says that

$$\tan\left(\frac{\theta}{3}\right) = \frac{\sin\theta}{2 + \cos\theta}.$$

A New Method of Trisecting Any Angle and of Constructing a Regular Pentagon with Ruler and Compasses, Together with Their Uses in Solving Other Geometric and Mathematical Problems

By H. A. F., 1904

H. A. F., superintendent of city schools, published this six-page pamphlet so that teachers of geometry could use his constructions. He had rediscovered the Archimedean trisection, and *Ruler* in his title could not be replaced with *Straightedge.*

Trisection

By N. W. G., 1985

Here is one of the rare trisections which does not start by bisecting the angle. In Figure 4.28, $|AC| = (1/3)|AB| = |BD|$. DE is perpendicular to BC and $|DE| = (1/2)|AB|$. Then bisect angle EOC to get the trisection.

The construction is not terribly accurate, but the error behaves in an unusual way: up from $0°$ to $9°$, then down until $18°$, then up until $69°$, then down to $90°$. No other error in this Budget has two bumps like this one. The error is also very large, exceeding $1°$ for many angles.

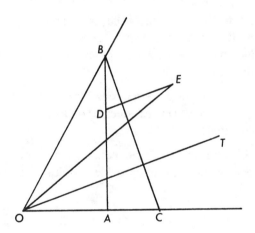

FIGURE 4.28

Portion of a Manuscript

By F. G., 1974

I have only two sheets by G., numbered page 44 and page 88, one of which gives a trisection. The construction (see Figure 4.29) is complicated. Erect a perpendicular to OA at O and draw an arc with center O and radius $|OB|$, determining C and D. Erect a perpendicular from D and draw another arc, with the same radius and center D, to get E and F. G is determined at the

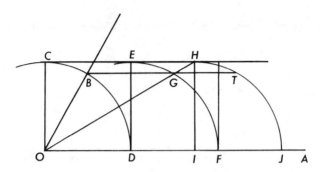

FIGURE 4.29

intersection of arc EF and an arc with center F and radius $|DF|$. Extend OG to the extension of CE to find H. Drop a perpendicular to OA to get I. Draw an arc, center I and radius $|OD|$, starting at J. T is the intersection of this arc and a line through B parallel to OA.

This may be a missimplification of the construction because it is not accurate at all—the maximum error for acute angles is $1°\ 45'$—but when I asked F. G. in what I thought was a polite way whether my interpretation was correct, his reply concluded

> Unless your University is prepared to publish my text, I do not think I should waste my time trying to educate you to understand what you obviously don't understand.

The handwriting was that of an old man. I had at second hand that G. had squared the circle also, but he would provide me with no details of that.

Letters

By P. G. G., 1964 and 1966

> G. wrote in the first letter that I have by him

> I've written to a couple of book publishers, wanting to incorporate my proof for trisecting an angle into high school Geometry textbooks. They're not interested.

> [One of them said] that I should give this to the world. I'm at a loss to understand why I should be asked to give my work and ideas, etc. to the world when authors, book and magazine publishers, etc. sell theirs. It doesn't quite make sense to me.

On the next page was the Archimedean trisection, rediscovered, with the same diagram that Archimedes undoubtedly scratched in the sands of Syracuse these 1,700 years ago. The second letter, sent two years later, shows the corrosive effect of time on a trisector.

> There certainly is no percentage in my submitting my proof to you because, one, you're not interested; you can't do it, so no one else can either. Secondly you've convinced yourself that the problem is impossible to solve, so you'd discard my proofs without being honest enough to look at them. . . .
>
> I've always believed that the problem was solvable, and I determined that I would find the answer, and since 1945, I have been determined that I would never again try to claim the job completed until I had irrefutable proof. I've gone at the problem with the positive approach that it can be done, and I've done it. It has taken a lot of sweat, a lot of spare time, and an awful lot of ridicule from people just like you, but I wouldn't quit until I found the answer.
>
> I've been trying to find someone with a name in the math field who has sense enough to realize what he could do with my proof in a new textbook. I've been trying to find someone who hasn't blinded himself with "proof" that it can't be done and who has at least enough interest to examine my proof before sneering that it can't be done. I haven't found that man yet, but I'll find him just as surely as I've found my proof.
>
> When [G.]'s Trisection of the angle finally does get the recognition it deserves, you can always tell your friends that you had the chance to help me but wouldn't do it. I don't have a name in the math field, I'm a nobody who never got beyond High School, but I'm smarter than you with all your degrees in math, because I can do something you can't do, trisect an angle.

The trisection is not the road to happiness.

To Trisect an Acute Angle

By C. S. G., in *School Science and Mathematics*, 1906

This construction and proof were no doubt printed to give the readers of the journal a chance to try their hands at finding the flaw, and shortly thereafter the journal printed a refutation by R. E. Moritz.

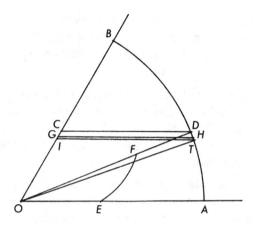

FIGURE 4.30

The construction is novel (see Figure 4.30). Take a convenient distance and draw CD parallel to OA. Draw OD. Draw an arc with center C and radius $|OC|$ to determine E and F. G is found by drawing an arc with center O and radius equal to the average of $|CF|$ and $|FD|$. GH is parallel to OA, and IT is parallel to GH and distant from it one-third of the distance from CD to GH. Though novel, the construction is not accurate, with its error rapidly creeping up toward 1° as angle BOA increases toward 90°.

The Cyclotome Absolute

By R. L. G., 1957

Only 100 copies of this privately published 200-page book were printed, and if I ever found a copy in a library I would be tempted to steal it. One of the lucky possessors of a copy wrote

> I can't send you a copy of the book because they are few and far between and I refuse to part with my copy for love or money. It is one of the glories of my library.

However the contents of the book are no mystery, since G. freely sent out parts of it:

> However, I am getting better results by using full-page abstracts rather than entire book.

and he distributed copies to anyone he could get to take them. He "started Project CYCLOTOME" while a freshman at the University of Vermont in 1905 and continued with it until his death in 1970. He was a teacher of mathematics and science in the schools of New York City for 38 years, retiring in 1953. He found mathematics difficult to teach for two reasons:

> (1)—He has had to teach much that is not necessarily true. (2)—He was not permitted to teach much that is necessarily true. Mathematics carries this straitjacket into other fields of science, but in a greatly diluted form.

His idea of the importance of his work was more grandiose than usual. He wrote in 1957

> This material may have unique importance due to the fact that the Russians had access to it in 1945. If it is valid, as I claim, and they took it seriously while others refused to look at it, and if they are developing it in secret we may be in serious trouble we do not even anticipate. They boast about a new and phenomenal interest and proficiency on the part of their youth in mathematics and technology; and ascribe this to reorganization, increased appropriations, govt. subsidy, guidance, etc. But we have good reason to believe that (barring mass mutation to a higher I. Q.) only discovery of new and simpler fundamental relationships, laws or formulae could accomplish such a miracle. What the Russians may have found out was that the ancient Babylonians made an error in their astronomy. It was a trivial error, but it was the seed from which grew momentous errors in the immature logic and mathematics of their time. Some of these errors are still bearing fruit of their kind, protected, not by logic, but by dogmas, doctrines, conventions, authorities, various kinds of brainwashing, and thought-control.

Some of the continual frustration that a crank must endure shows through. The error made by the Babylonians was to divide the circle into 360 parts. They should have divided it into 120 parts.

G. tended towards belief in conspiracies:

> The new material in this book is entirely original, although any or all of it may have been formulated and offered for publication many times only to be suppressed or "hushed up" in the interest of great dignity in high places, and in order to protect the private and public intellectual and financial investment in Ph. Ds. A grotesque and fantastic world-wide cartel and censorship based on very narrow spe-

cialization and "professional ethics" in the intellectual realm permits
no research or investigation for credit towards degrees of any rank in
the territory in which this new material has been found to exist.

Mathematicians who have dealt with trisectors have seen only one side of
a two-sided transaction. Here is how the other side looks:

> More than twenty years ago the writer started efforts to get a small
> amount of material before several mathematical societies, but has not
> succeeded yet. ... A more or less standardized procedure "turns the
> trick" very effectively and neatly, as follows:

> When new material ... is offered by a non-member of the society it is
> at once labeled "trisection" and rejected on the ground that only the
> work of members can be considered. If a member tries to reach the
> membership through the society's publications, his work is rejected
> on the ground that it first must be presented on the convention plat-
> form. If he asks for time on the platform, dissuasion is used and he is
> reminded of the heresy of his position.

G. was a member of the American Mathematical Society until he quit in dis-
gust. He then inquired about joining the Mathematical Association of America. Members of the MAA may think that anyone who can pay the dues can
become a member, but not so. At least it was not so in 1944, when the Secretary-
Treasurer of the MAA wrote to G. saying, according to G.,

> I am somewhat afraid that membership in our organization would not
> be more satisfactory to you than your membership in the American
> Mathematical Society from which, I understand, you have resigned.

"Of course," G. wrote, "I never applied for membership in the Mathematical
Association of America."

Returning to G.'s view of how he was treated:

> If persistent, he is given, with great reluctance, a so-called "referee"
> located far away and hostile to the subject.

> Then a "comedy of errors", misunderstandings, and intellectual abuse
> begins and drags on from months into years of nonsense. One can-
> not hurry his "referee", so years pass and his existence is forgotten.
> In the meantime, if any contact with the "referee" is maintained, he
> (the referee), even though beaten again and again "slides under each
> fence, clears his conscience, and proceeds blithely to the next truth
> only to violate that in turn." (Quotation is from a description of Don
> Quixote in mathematics, by Col. Yates [the author of *The Trisection*

Problem].) Over and over again the "referee" declares himself victor on all points that have been discussed but not disposed of, and sets new and "final" requirements, often impertinent, that must be met before a favorable report on the subject can be rendered. It is the writer's experience that if contact is maintained with the "referee" he will endure the embarrassment and unavoidable humiliation of repeated defeat from one to two years, depending on his disposition and the skill with which one is able to spare his feelings; then he will resort to summary action in great dignity and dudgeon and declare the case closed in glorious victory for himself; and there is no appeal except for another "referee", who, very probably, will live still farther away, and would consider himself disgraced if he were to yield where the first referee had stood firm in defense of the most sacred icon of the faith.

Frustration!

The deeper we go into mathematics the deeper we find ourselves in human behavior. In professional societies mediocrity climbs to exalted position, by methods that are understood very well when encountered in politics, and then uses that position as a platform for heroic dramatics, which, for lack of other substance must consist chiefly of smiting to the death any accomplishment offered from below that might attract the eyes of the audience for so much as a moment.

After giving up on the AMS, G. appealed to the public at large, sending out one hundred and ninety-seven letters, to people as varied as Albert Einstein, Dwight Eisenhower, Secretary of the Navy James Forrestal, the president of the National Geographic Society, the editor of the *New York Times,* Franklin Roosevelt, the astronomer Harlow Shapley, and others, many with mathematical connections but several without. He got, he says, one hundred and thirty-three replies!

Not all of the replies were entirely serious:

One mathematician in cleric garb volunteered to examine the work. All went well until he found that the Galois theory was being challenged. Then he rejected everything on the ground that, if his memory did not fail him, Galois had been made a saint: hence his work was not subject to such a review.

Clever! G. looked into the matter and found that Aristede Galois was a saint, but Evariste was not. By then the priest had made his escape.

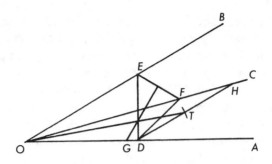

FIGURE 4.31

G.'s trisection is illustrated for a 30° angle since it is easier to see how the construction goes and because it quickly becomes very inaccurate for larger angles. In Figure 4.31, *OC* bisects angle *BOA* and *D* bisects *OA*. Angle *EDA* is three times angle *BOA*, and *DF* is its bisector. *G* is on the perpendicular bisector of *EF*, *DH* is parallel to *OB*, and *T* is |*EG*| from *G*. The error is less than 2′ for angles less than 30°.

The Mathematical Atom

By J. J. G., 1934

I have only an advertisement for this trisection, so I have no idea how good the construction is. The advertisement includes an illustration of the trisection diagram but no explanation. It may be possible to reconstruct the steps of the trisection from the diagram, so Figure 4.32 has been included if you want to try.

This is the author who thought that the trisection might make transmutation of elements possible. He had other applications in mind as well:

> It is well within the range of possibility, and even probability, that with the discovery, as a by-product of TRISECTION, of a new GEO-METRIC UNIT, and of a number of new, so-called higher curves . . . [science will have] an important aid in establishing and setting forth more clearly the laws and principles that are at the back of SOLAR RADIATION as well as REVOLUTORY and ROTATORY power and motion.

The complete work cost $1.50, rather too much to pay in 1934 even though

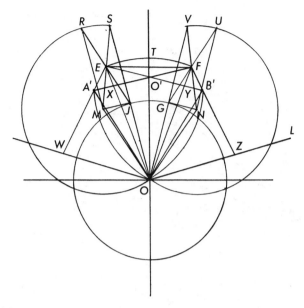

FIGURE 4.32

Returns, great or small, will enable the author's confreres at the Franciscan Friaries up and down the land to continue to be God's almoner to the poor, helping to tide them over until such time when happily, like the angle, the good things of life shall have been more evenly divided.

Rarely does a trisector write with such grace! Had the project been almost anything except the trisection, I would have wished Brother J. good luck and Godspeed.

Construction

By A. C. H., 1983

H. sent me a trisection with "N. B.: This method fails in the range of angle 1–60." It certainly did, and it also failed for larger angles, with errors sometimes exceeding 1°. When I pointed this out to H., almost by return mail (as return mail goes between India and Indiana) came the following construction (Figure 4.33): $|BO| = |OA|$, OC bisects angle BOA, $|OD| = |OA|$, E bisects DA, T is

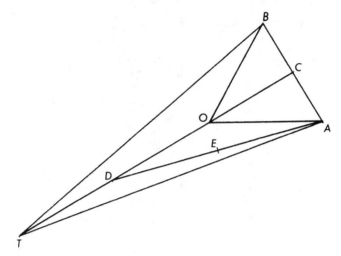

FIGURE 4.33

on OC extended with $|DT| = |DE|$, and angle BTA is the trisection of angle BOA. It is not bad for quick work, with errors less than $6'$ for acute angles, and it is equivalent to asserting that

$$\tan\left(\frac{\theta}{3}\right) = \frac{\sin\theta}{1 + 2\cos(\theta/2) + \cos(\theta/4)}.$$

The Trisection of Any Rectilineal Angle

By G. H., 1911

H. gave his occupation as "contractor" on the title page of this clothbound 20-page booklet. He wrote

> In the presentation of this problem there will be some repetition which may be open to objection as unnecessary detail; but it is considered preferable to err in that way than risk the omission of something that ought to be stated.

It is too bad contractors can't write that way any more. It is too bad that hardly anyone can write that way anymore! The bulk of the text is given over to a long, long demonstration of the correctness of the trisection; I could not face the labor of plowing through it.

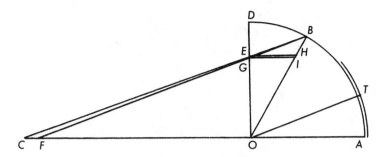

FIGURE 4.34

The construction gives a good approximation, with maximum error of 3′ for acute angles, occurring almost right at the 60° angle that trisectors always attack. In Figure 4.34, $|OC| = 2|OA|$, OD is perpendicular to AC, and E is the intersection of BC and OD. F is on OC, $2|OA|$ from E, and G is the intersection of FB and OD. EH and GI are parallel to AC, and T is on an arc with center O and radius $|OA| + |HI|$ at that distance from E. H. had published an earlier trisection—I suspect that the little distance $|HI|$ was the substance of the revision.

Untitled Trisection

By B. H., date unknown

In Figure 4.35, OC is the usual bisector, $|OD| = 2|OC|$, E bisects OA, and T is the intersection of AC and ED. This locates the

$$\tan\left(\frac{\theta}{3}\right) = \frac{2\sin(\theta/2)}{1 + 2\cos(\theta/2)}$$

trisection point which comes up several times in this Budget. It is evidently the easiest trisection point to find.

Manuscript

By C. H., 1973

Here, simplified from H.'s manuscript, is one of the simplest possible trisections. In Figure 4.36, construct the equilateral triangle ABC and bisect AC to get T. That's all!

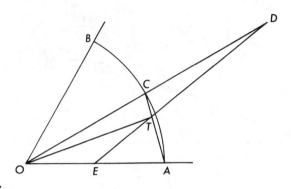

FIGURE 4.35

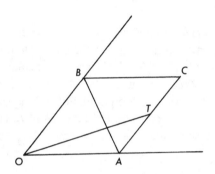

FIGURE 4.36

It is equivalent to saying that

$$\tan\left(\frac{\theta}{3}\right) = \frac{\sin(\theta/2)\sin(\theta/2 + \pi/6)}{1 + \sin(\theta/2)\cos(\theta/2) + \pi/6}.$$

but they are not equal and the maximum error for acute angles exceeds 1°. There is a one-page "proof" which rests on an unsupported assumption, very easy to spot. If the author showed his construction around, he would quickly realize that it needed improvement. I do not know whether he ever pursued the trisection further.

Trisection

By M. G. G., 1961

G. claimed that more than one hundred people had the opportunity to examine his trisection and none had been able to find any error in it. It is curious that with such wide circulation G. never had his construction printed. It is also curious how his mind worked; I found an error in his trisection, and I am sure that many other people did too. What he meant must have been that no one had been able to convince him that he could be in error.

The construction is complex and makes one wonder how it could have been developed. In Figure 4.37, OC is the bisector of angle BOA and D is its intersection with AB. OE bisects angle COA. F is on OE extended, where $|OA| = |AF|$. Points G, H and I are very close together; in Figure 4.38, angle

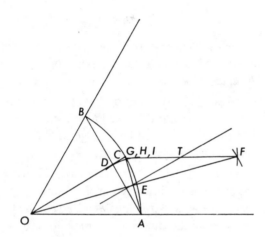

FIGURE 4.37

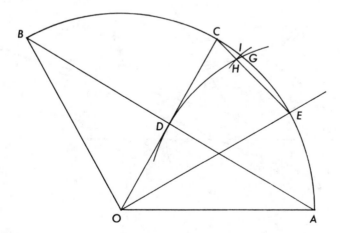

FIGURE 4.38

BOA is bigger so as to increase their separation. *G* is the intersection of arc *AB* with an arc with center *A* and radius |*AD*|, and *H* is the intersection of this arc with *CE*. *I* is the intersection of arc *AB* with an arc with center *E* and radius |*EH*|. *T* is then the intersection of *IF* and the perpendicular bisector of *AD*. The construction is very accurate, with maximum error for acute angles no more than 14″. Of course, if it were going to be done with a real straightedge and compass, a very big diagram would be needed.

Trisection

By V. H., 1908

A professor of mathematics at Auckland University College, writing on the three problems (trisection, quadrature, duplication) in the *Transactions of the New Zealand Institute,* 1908:

> Occasionally, however, one or other of these problems takes posses-
> sion of some person unaware of these investigations [proofs of im-
> possibility], and with only a slight knowledge of mathematics. I was
> recently approached by Mr. [V. H.], of this city, ...

With this construction (Figure 4.39): draw the circle and bisect the angle to get *CD*. Bisect again to get *F*, and the intersection of *DF* and *EC* gives *T*. A simple construction, and accurate, as constructions which go behind the angle tend to be. There is something about the stately march of points around circles behind

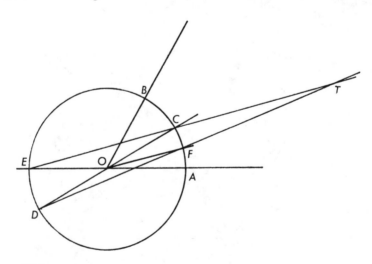

FIGURE 4.39

angles that keeps constructions in hand and prevents them from exploding into raving craziness. That last sentence is an attempt to communicate a mystical insight into trisections, difficult to do in words. The error for acute angles is less than 3′ and is less than 20′ at 180°. Also, the angles determined are always a little bit less than the true trisection: if only we could find a way to push T up, just a little bit. . . .

We were not told whether H. was satisfied with finding only an approximation. Being a New Zealander, and living in a time when respect for authority was more common than it is today, he probably was.

A Garden of Geometrical Roses

By T. H., 1727

This is the English translation of *Rosetum Geometricum,* published in 1671. H. (1588–1679) was a well-known political philosopher who rates eight pages in the *Dictionary of National Biography* and who caused a "fermentation in English thought not surpassed until the advent of Darwinism." As Aubrey tells us in his *Brief Lives,* H. became interested in mathematics when he picked up a copy of Euclid; opened it to Proposition 47; said, "By God, this is impossible!" and then went to the beginning and worked his way through Propositions 1–46. H. also squared the circle and duplicated the cube. H. did not know that these feats were impossible, since the proofs of impossibility were not accomplished until the nineteenth century, so he is not like the other trisectors in this Budget. I was thus tempted to reveal his name, but rules are rules. Perhaps you can guess, with the aid of the hint that one of his other works was *Leviathan.*

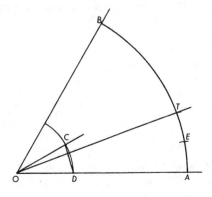

FIGURE 4.40

In Figure 4.40, $|OA| = 3|OD|$, OC is the bisector, and $|AE| = |ET| = |CD|$. The simple construction is equivalent to

$$\cos\left(\frac{\theta}{6}\right) = 1 - \frac{2}{9}\sin^2(\theta/4)$$

and trisects acute angles with errors less than $42'$, though the error at $60°$ is only $12'$.

A New Construction for Approximating the Trisection of an Angle Using Only a Compass and an Unmarked Straight-Edge

By T. J., 1971

As can be seen from the title of this neatly printed nine-page work, the author was no trisector since he knew that his construction could only be approximate. He did not know how easily such constructions could be found nor of how little importance they are geometrically. But the connoisseur of trisections notes that this is the only construction in this Budget which uses the bisector of the angle to be trisected and duplication of angles. It is thus a rare and valuable find, even as would be an unusual stamp, book, butterfly, or piece of barbed wire to collectors of such things.

In Figure 4.41, angle COB is one-half of angle BOA, and angle ACD is the duplicate of angle ABO. Angle CAE equals angle OBA, and the intersection of CD and AE determines F. Extending BF gives C, and then T is found by making $|AT| = |GE|$. The error increases steadily: it is $12'$ at $60°$ and a large $43'$ at $90°$.

Trisecting

By M. E. J., 1972

This construction, with proof, was sent to the *Bulletin* of the American Mathematical Society in 1974. What makes it unusual is that the author is female: women, as was pointed out earlier, trisect hardly at all.

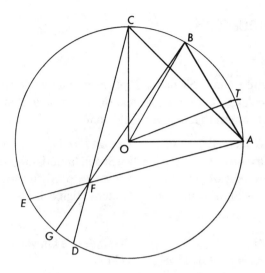

FIGURE 4.41

In Figure 4.42, OC is the bisector of angle BOA and $|CD| = |OC|$. The circle with center at D has radius $|OA|$. OE is perpendicular to OA and $|OE| = 2|OA|$. T is the intersection of FG, which is tangent to the two arcs, and a line through E parallel to OA. The error is consistent with the simplicity of the construction.

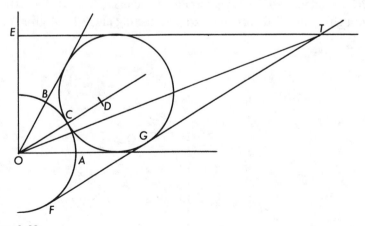

FIGURE 4.42

Untitled Manuscript

By A. L. J., 1970

J. was careful to have his manuscript signed and witnessed so that after copies were circulated no one could steal the construction. Since he lived in Michigan, it was natural that copies should go to

Cass Tech High	Detroit *Free Press*
Detroit *News*	Michigan State University
University of Detroit	University of Michigan
Wayne State University	Eastern High School

but why did the other copies go to

MIT	St. Andrew's University
Scientific American	UCLA
University of Dundee	

of all possible places? One other copy was sent to *Mechanix Illustrated*: the 1966 publication in that magazine of H. C.'s trisection bears fruit!

The construction is clear, the error is easy to spot (it is in step 35 of a 38-step proof), and the writing is literate: I hope that J. had the intelligence and inner resources to shrug off his trisection defeat and forget about it. The angle to be trisected is never bisected, the common first step in trisections, and almost all of the construction goes on behind the angle, also not usual. Draw circle ABC (Figure 4.43) and make $|CD| = |CE| = |OA|$. Drop a perpendicular from AC at C and let F be its intersection with AD extended. Draw FE and its perpendicular bisector, GH. The arc EF has center H, and I is its intersection with BC extended. Draw AI to determine J and make $|AJ| = |JK|$. Then the line through K and C determines L, and bisecting angle LOA gives T. How

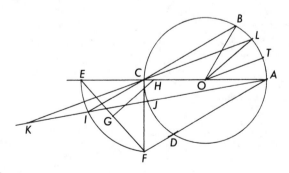

FIGURE 4.43

could anyone have thought of that? The error is less than 10′ for acute angles and less than 20′ for angles up to 180°.

Trisection of Any Angle with Compasses and Straightedge

By C. J., 1961

J., a holder of the M.D. degree, started his mimeographed trisection with

The ancients believed that it was impossible to trisect any given angle (the right angle excepted) by means of compasses and straightedge alone. This belief has persisted to the present time.

It is strange that people think that the truths of mathematics are matters of opinion.

J. offered no proof for his trisection, and I would guess that he could have been convinced that he had made only an approximation. In Figure 4.44, AB is perpendicular to OA, and OC, AD, and BE are angle bisectors. Angle ADF is twice angle ACE, and FG is parallel to EC. Angle EAT is 30° and T lies on FG. The construction trisects the 45° angle precisely and has no error greater than 11′ for angles up to 72°, but it goes to pieces rapidly after that.

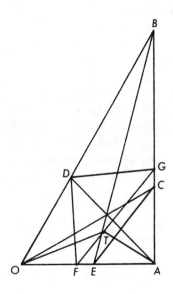

FIGURE 4.44

Untitled Trisection

By W. J. J., 1973

This is an incredibly complicated construction, on blueprint paper, whose description takes up three typewritten pages. But it determines the same point as the other trisections that claim that

$$\tan\left(\frac{\theta}{3}\right) = \frac{2\sin(\theta/2)}{1 + 2\cos(\theta/2)}$$

so no figure is needed, since the others are simpler. J. (in his eighties at the time) was able to hold simultaneously in his mind the truth of a proposition and the truth of its negation:

> After a thorough reading of the mathematical disproof that an angle cannot be trisected geometrically and being aware of the governing rule for pure trisection, I have this to say: the solution, as it stands, will divide any given angle into its three equal component angles, yet it might not be a true trisection.

Trisection of Angles

By K. O. J., 1973

The construction, a good deal simplified from J.'s version, is not hard (Figure 4.45). Draw arc AB, erect a perpendicular to OA at O, and draw a semicircle with CO as diameter. Drop a perpendicular from B to OA at D (it is tangent to the semicircle because angle BOA is 60°; it will not happen in general) and transfer the distance $|DA|$ to the vertical line as OE. OF has length one-fourth $|OE|$ and FT is parallel to OA; T is the intersection with the semicircle.

The construction amounts to claiming that

$$\sin\left(\frac{\theta}{3}\right) = \frac{\sqrt{2}}{2}\sin(\theta/2)$$

which is true when θ is 0° or 90° but is not very true elsewhere: the error is almost 44′ near 55°.

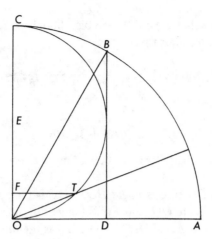

FIGURE 4.45

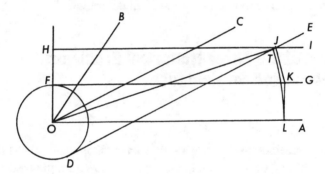

FIGURE 4.46

Trisecting Any Angle

By J. J., 1940

This small pamphlet, neatly printed with blue ink, shows traces of the evolution of a trisection. In Figure 4.46, point J gives an approximate trisection that has been found many times (see the next two trisections and the one before last): it is the one that amounts to saying that

$$\tan\left(\frac{\theta}{3}\right) = \frac{2\sin(\theta/2)}{1 + 2\cos(\theta/2)}$$

It always gives angles that are a little too big, though, so J. adjusted J slightly; his adjusted trisection claims that

$$\tan\left(\frac{\theta}{3}\right) = \frac{D - \cos(\theta/2)}{D\cos(\theta/2) + \sin(\theta/2)}$$

where

$$D = \sqrt{2\cos^2(\theta/2) + 4\cos(\theta/2) + 3},$$

and the accuracy is not bad, the error being only $3'$ at $60°$ and less than $10'$ for acute angles.

In Figure 4.46, OC bisects angle BOA, DE is tangent to the circle with center O and is parallel to OC, and FG is tangent to the circle and parallel to OA. HI is parallel to OA and $|HF| = |FO|$. J is the intersection of HI and DE. An arc, center O and radius $|OJ|$, determines K. KL is perpendicular to OA and an arc with center O and radius $|OL|$ determines T.

Solutions of the Three Historical Problems by Compass and Straightedge

By D. J. J., 1962

This 32-page booklet has on its title page "Copyright 1962-1963-1965-1968-1971-1972," showing the frequent revisions. Even so, the trisection is by assertion. That is, in his "proof," the author merely described his construction and concluded with "Angles in ECF is the required angle." In common with most trisectors, the author had no knowledge of trigonometry and evidently not much of algebra either, because in solving the problem of duplicating the cube he wrote

> A given cube can be doubled in volume if the area of the sides of a cube is doubled. The volume of such a cube is not 2 times the volume of the original cube, but $2\sqrt{2}$ times, as any student of high-school algebra ought to be able to calculate.

I wrote to J. about his trisection, but, unlike other trisectors who are happy to find a mathematician to write to, he never answered. I thought that he might have lost interest in the trisection, but he had no such luck: I have been sent a copy of his work, which was published by a vanity press in 1982. The contents do not seem to be any different from those of his privately printed booklet of 20 years earlier.

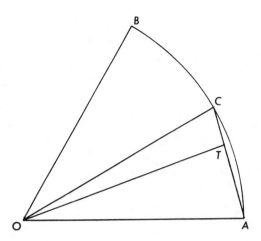

FIGURE 4.47

In Figure 4.47, OC bisects angle BOA and CT trisects the segment CA. This finds the same point as J in the preceding trisection, and the same construction appears in Steinhaus's *Mathematical Snapshots* as an approximate trisection. Steinhaus wrote "A lawyer, Mr. Rappaport, [is] the author of this trisection." The error is 6′ at 60° and increases steadily to 22′ at 90°.

The General Angle Trisected

By D. R. J., 1960

> J. wrote

> I know from experience that high-brass in the field will not read notes from amateur mathematicians on the trisection subject. Those fellows have never experienced and will probably never know how sweet it is to learn from a child. The lesser lights are more sociable about it.

His trisection experience was long:

> So on the streetcar on my way home from work [in 1937] I thought of a way to try. .... After I got home I put on paper what had turned over in my mind and there it was.

What it was is in Figure 4.48, where OC bisects angle BOA, angle CDT is one-fourth angle BOA, and T is located by making $|DT| = 2|OD|$. The approximation is good, with maximum error a mere 2′ 36″ for acute angles. It

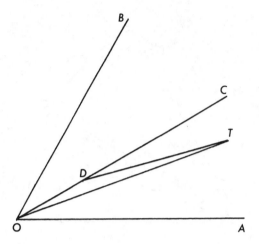

FIGURE 4.48

is equivalent to asserting that

$$\tan\left(\frac{\theta}{3}\right) = \frac{2\sin(\theta/4) + \sin(\theta/2)}{2\cos(\theta/4) + \cos(\theta/2)}.$$

Supplement to [J.]'s Geometrical Division and Measurement of Arcs and Angles

By N. J., 1900

This is a two-page supplement to a larger work that I have not seen. The construction determines the same point as the preceding trisection by D. R. J. It is clearly of mystical significance that J. is a common first letter of last names (Johnson, Jones, and so on) and that this is a common trisection.

In Figure 4.49, OC bisects angle BOA and T is the intersection of BC and the perpendicular to OA erected at A.

Trisection of an Angle with Ruler and Compass

By R. J., 1986

J. said that he had sent his paper to many universities. He also made the incredible claim

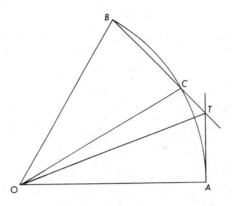

FIGURE 4.49

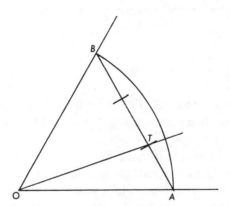

FIGURE 4.50

> More than 50 professors of Mathematics (many of them are Ph.D.s) have evaluated my paper and approved it.

What could he mean? Was he lying, or was it that anything short of physical violence counted as approval? Having disposed of the trisection, he wrote that he is now proving Euclid's fifth postulate, proving Fermat's last theorem, squaring the circle, duplicating the cube, and settling the four color problem, probably unaware that the last has already been taken care of. He said also that he is the creator of a new branch of mathematics, mathematical birthology.

After reading his trisection, I went back and read it again, several times, very carefully, since I thought I must be missing something. But I was not: J.'s trisection (Figure 4.50) simplifies to that construction that has been thought of thousands of times by high schoolers and, after reflection, discarded thousands of times: strike the arc, draw AB, trisect it, and there you are. There you are

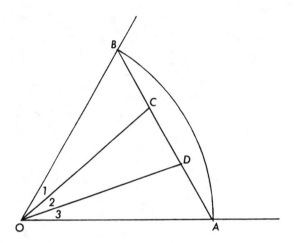

FIGURE **4.51**

with a construction so inaccurate that the eye should be able to see that it is wrong, and so simple that geometry tells you right away that it is wrong. The error at 60° is almost 1°, and it increases.

There are many ways to see that this construction does not trisect the angle, including drawing an obtuse angle and using your eyes. Here is the geometrical way I like best. In Figure 4.51, suppose that the trisected segment *BA* trisects the angle, so that angle 1 = angle 2 = angle 3. Look at triangle *BOD*. The segment *OC* bisects angle *BOD* and passes through the midpoint of the opposite side. But that happens only in isosceles triangles! Therefore $|OB| = |OD|$. But *OB* is longer than *OD*, so the assumption that the angle was trisected has got to go.

The Paradox of the Angle Trisection

By K. O. L., 1987

This trisector, who submitted his construction to *Mathematics Magazine*, was much as many others:

> The author has not been able to disprove Prof. F. Gauss nor Mr. P. L. Wantzel; however, neither has he been able to find a flaw in his own method. Therefore the problem of the angle trisection will have to be

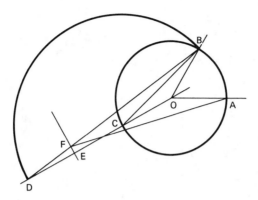

FIGURE 4.52

called a paradox as long as your mathematicians will not be able to find a flaw in the author's method or to dispute Prof. F. Gauss's proof.

(By "Prof. F. Gauss" he meant Carl Friedrich Gauss.) His construction is unusual in that the trisected angle does not have its vertex at the vertex of the original angle. In Figure 4.52, bisect the angle, draw the circle ABC, draw an arc with center C and radius BC, bisect CD to find E, and erect a perpendicular at E to find F. Angle BFA is the trisection.

L. was aiming at the Archimedean trisection point, and he was not far off: $60°$ was trisected to within $46''$, and the maximum error for acute angles is $2'\ 36''$.

A Proposition to Trisect an Arbitrary Angle

By R. A. L., 1978

L. wrote that

the search for the trisection of the general angle continues unabated by mathematicians and laymen alike.

He was only half right: mathematicians long ago abated. He went on

I offer no criticism of previously attempted solutions of the Trisection of the general angle, and have the greatest respect for the position held by those who consider it a closed subject. Further, I make no dogmatic claim for my proposed construction, but merely state that it appears to trisect the angle.

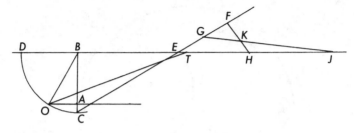

FIGURE 4.53

But he was hopeful enough that he went to the trouble of having printed and copyrighted a booklet, complete with National Library of Australia Card Number and an International Standard Book Number.

The construction aims at the Archimedean trisection point and is more complicated than most. In Figure 4.53, BC is perpendicular to OA and BD is parallel to OA, with $|BC| = |BD| = |BO|$. E is on BD extended, at a distance $2|BO|$ from C, and F is another $|BO|$ units along. G is $|OA|$ back from F. $|BH| = 3|BO|$, and J is $|CD|$ further. K is the intersection of FH and CJ. Finally, T is on BD extended, $|BO|$ from K.

For such elaboration, the construction is not very accurate, the error being $31'$ at $60°$ and as much as $55'$ for acute angles. There is the possibility that I did not follow it properly.

The Quadrature of the Circle Proved, with Diagrams and Numerical Formulas for Practical Purposes

By H. L., 1847

L. was mainly concerned with squaring the circle, which he thought would be helpful in determining longitude at sea, a problem which in 1847 had no practical solution. It was solved when technology progressed enough so that accurate chronometers could be made, and squaring the circle had nothing to do with it. His introduction claims

> The following Mathematical Diagrams will present a geometrical so-
> lution of the four problems hitherto unattained: First, The Quadra-
> ture, or squaring the Circle; Second, The relation between the cir-

cumference of a Circle and its diameter; Third, Trisecting any given angle; Fourth, Trisecting any given straight line.

That he thought that the problem of trisecting a line segment was unsolved indicates the level of his mathematical training. (For the way it is done, see Figure 4.54: to trisect AB, slap down any line AC, step off three equal lengths with your compass, and draw three parallel lines.) In Figure 4.55, C is constructed so that triangle ABC is equilateral and D is the bisector of BC. The distance from D to T is the same as the distance from D to C. The construction is the Steinhaus et al. one once again.

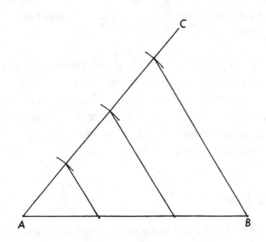

FIGURE 4.54

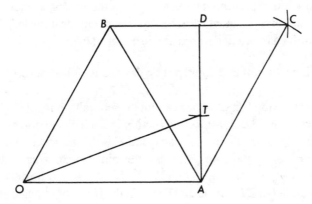

FIGURE 4.55

Trisection of the 120 Degree Angle

By G. W. L., 1973

This is a hardbound book, 57 pages long, published by a vanity press. I doubt that L. recouped very much of what he had to pay to have it printed, since I cannot imagine that anyone besides me bought a copy. There is no market whatsoever for crank mathematics: mathematicians will not pay for such things, nor will other cranks, and the general public is not interested. I have a copy of a crank physics book, privately printed, with "Price $150" on it in big letters. Its author might just as well have made it $1,500, since his income would be the same. Paying for publication was a great waste of L.'s resources, not to mention the trees cut down to make the paper on which it was printed.

His introduction is curious. It includes

> Each of the three problems has been proven to be impossible of solution when carried out within the restrictions of the allowable instruments, the straightedge and compass. ... The impossibility has been demonstrated by the experts. ... The trisection, according to the experts, results in an irreducible cubic equation, and again in the language of the experts, is impossible of solution with straightedge and compass.

And yet he went ahead!

> Many times, the author has been constrained to shrink from further effort by subtle innuendo and the language of the experts. Because of this the solution presented here, along with the proofs, is offered not in stubborn disagreement with the experts, but in the knowledge that a old truth has been uncovered and brought to light.

It seems as if he knew that it could not be done, but he had nevertheless done it.

The construction (Figure 4.56) cannot be generalized to all angles. Angle BOA is, as usual, $60°$ but for this construction it can have no other value. Arcs BC, CD, and AE also subtend $60°$ each, and F is the intersection of DE with BO extended. FG is parallel to OA; G is determined by drawing an arc, center A and radius $|OA|$. Extending AG gives H, and T is found by drawing a line through H parallel to BF. Angle BOT is then the trisector of angle BOA, very nearly: the error is only about $2'\ 30''$.

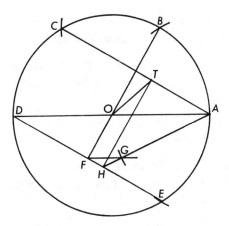

FIGURE 4.56

Challenging and Solving the "3 Impossibles": Trisecting the Angle, Squaring the Circle, Doubling the Cube, with Compass and Unmarked Ruler Only!

By M. L. in Collaboration with J. X. W., 1961

This tireless pair (L. was the trisector and W. the promoter) solicited endorsements; they induced their representative in Congress to insert notice of their accomplishments in the *Congressional Record*; they asserted that their solutions were taught in more than fifty schools and colleges; they printed and offered for sale several versions of their constructions, energetically advertising them; and they hoped to make their fortunes selling [L.] Ratio Calipers:

> Easy to handle, light in construction with unlimited uses! Made of the highest quality tool steel and the finest aluminum, anodized in beautiful gold finish to prevent tarnishing. Each [L.] RATIO CALIPER is highly calibrated for accuracy and precision. Made to render years of useful service!

It was enough to make me regret not having one, but not enough to make me spend my money to get one.

Another advertisement:

> Now! The book the world has waited for … Prove it to yourself! Learn the amazing [L.] SOLUTIONS to the problems that have baf-

fled experts for Centuries! Simple Constructions! Logically Sound!
Fully Described! Clearly & Profusely Illustrated with Proofs Galore!

"Proofs Galore!" was a mistake since many people remember their tenth grade
geometry proofs with something much less than fondness. Yet another advertisement was headed

> [L.] Solutions Win Top Awards at the 1959 Chicago City and Illinois
> State High School Science Fairs!

which is hard to believe, and dismaying if true.

Of course, they tried always to get publicity, favorable or not. A story in
the Los Angeles *Herald-Examiner* ended

> "I can do it," says [L.]. "If you can," retort local university mathemati-
> cians, "you can also prove two and two are five." That, undoubtedly,
> is the next problem.

It would all be funny if it were not so pitiable. This deluded pair, not only
making themselves ridiculous but eagerly seeking out opportunities to make
themselves ridiculous! I hope that not many in those more than fifty schools
and colleges became similarly deluded.

L. was aware that the trisection had been proved to be impossible, but he
was not clear on what *proof* and *impossible* mean in mathematics. On impossibility:

> The author resents the negative implications since, if everyone were
> to accept statements as valid, there would be very little progress.

On proof:

> A mathematical proof is merely an established approximation, indi-
> cating a limitation of errors to a minimum applicable to each on hand
> to be solved, and from a point, or points, of reference as they appear.

Read that again, carefully. It is impossible to parody trisectors because it is
impossible for an ordinary mind to produce such inimitable nonsense as the
preceding. It is a skill that not everyone has.

The L. trisection is one of the very few to appear in a scientific journal.
In 1961, the editor of *School Science and Mathematics,* who I think was not a
mathematician, wrote under the heading

Challenging the Impossible

on the subject of authors attacking "problems long thought impossible of
solution":

In fact, one of the supposedly outstanding mathematicians in the United States commented about one effort in his letter as follows:

"No, I haven't seen any of their material and I won't waste my time on it. I know they are wrong."

Such a statement is a reversion to barbarism. The Editor is certain that any scientist worth his salt is eager to examine a new idea for a perpetual motion machine to see how the hapless inventor sought to "beat the game." He is reminded also of the indivisible atom, the earth as the center of the Universe and the recanting of Galileo, and the conflicting views of inheritance.

So, the editor decided to print works of such authors from time to time. I think he later realized more fully the difference between atomic theory, evolution, and cosmology on the one hand and mathematics on the other. Science uses induction, while mathematics uses deduction. The Law of Gravity may be repealed tomorrow, but the Pythagorean Theorem is true forever. In any event, I did not find much crank mathematics in later issues of the journal. But there L. and W. are, reprinting diagrams from their pamphlet and trisecting by assertion. As might be expected from such a pair of go-getting, live-wire Americans (L. was born and raised in Europe and W. was Chinese-American), their ultimate argument was practical:

The artists and artisans of today who are familiar with the [L.] System extol its virtues in solving their problems of proportions and spatial harmony for practical use. As to the academic, hair-splitting proofs of the validity of the system, they couldn't care less! The important thing to them is that it works!

The construction is original in using neither the chord of the angle to be bisected nor its bisector. In Figure 4.57, arc BC ("the Grand Arc," L. called it) has center A and radius $|AB|$. D is the intersection of arc BC with an arc with center B and radius $|AB|$. Arc DE has center O and radius $|OD|$. F is determined so that C, E, and F are vertices of an equilateral triangle; the line from B to F intersects the arc DE at C. T is found by striking an arc with center G and radius $|GA|$. The construction has a maximum error of $30'$ for acute angles, not especially impressive.

I have wondered what happened to L. and W. and what effect the trisection had on their lives. Trisectors tend to drop out of sight.

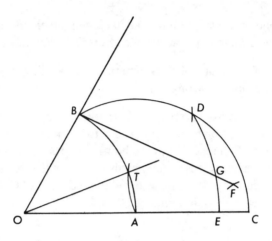

FIGURE 4.57

The Trisection of the Angle: Being a Problem in Geometry That Has Baffled the Efforts of Mathematicians Up to the Present Day, Now Solved for the First Time

By J. A. L., B. A., M. D., 1890

Concerning the problems of the quadrature of the circle, the duplication of the cube, and the trisection of the angle, L. wrote

> Many attempts have been made at their solution in every age and in every nation, but, up to the present time, the result has been considered unsatisfactory, so that now-a-days the solution of these questions has come to be looked upon as impossible. But why they should be impossible no one has attempted to explain. . . . I say no one, and here it may be thought that I am ignorant of the writings of Montucla, De Morgan and others, but I am not. . . . I therefore selected some years ago (it matters not how many), the problem known as the Trisection of the Angle simply because I could not see any reason why it should be looked upon as impossible.

By 1890 it had been completely explained why the trisection was impossible, and it is strange that an author who claimed to be familiar with some of the literature would not know that. I suspect that he began on his trisection work, put quite a bit of effort into it, and only then began to look around to see what

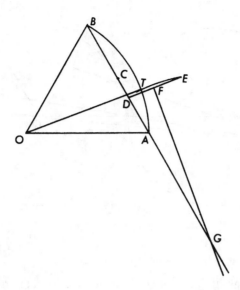

FIGURE 4.58

others had written on the subject. If you have enough invested in a project, it is possible for your eye to skip over those parts of books that say that the project is impossible.

Be that as it may, in Figure 4.58, C is the midpoint of BA and E is constructed equidistant from C and A. Another trisector stopped there, but L. saw that E was not a satisfactory trisection point. D is one-third of the way from A to B, and F is the midpoint of DE. G is the intersection of BA extended with the perpendicular to DE erected at F. T is on arc AB, $|GE|$ away from G. The error is never more than $9'$ for acute angles, rather good, and it is at most $19'$ for any angle from $0°$ to $180°$. But since G is the intersection of two nearly parallel lines, it would be hard to get this accuracy in a drawing.

Trisection

By R. M., 1986

M. wrote

I've been trying to get my trisection published for over a year, but the great ghost of Gauss is too strong for the logic of Euclid . . . one group felt it could get the thing into print. Unfortunately, after someone on

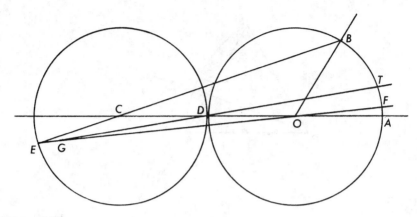

FIGURE 4.59

the staff thought they had discovered an error (misunderstanding the nature of the rhombus) publication promises were withdrawn, and I was left to gnash my tricuspids.

The construction uses two circles (Figure 4.59); the one with center C has radius $|OA|$. BC is extended to get E, and EO is extended to F. G is on EF, intersecting a circle with center D and radius $|DB|$. Finally, GD extended gives T.

Constructions behind the angle *always* give good results, and this one is accurate within $1'$ for angles up to $81°$, and the error at $60°$ is a mere $5.9''$. Impressive! However, it falls apart rapidly after $135°$.

Geometry

By D. E. M., 1964, 1971

This is a four-page pamphlet with handwritten additions. The writing is shaky. The construction is pleasing to the eye and is easily made. In Figure 4.60, angle BOA is bisected and the bisector is extended backward: $|OC|$ is twice $|OB|$. DE is the perpendicular bisector of OC, and F and G are intersections of DE with CB and CA, respectively. T is on arc AB at a distance of $|FG|$ from A. Here is another construction which goes behind the angle, and it is another one which is accurate over a wide range. The error increases to $6'$ at $60°$, gets up to $17'$ at $90°$, and hits a maximum of $40'$ near $160°$, after which it decreases

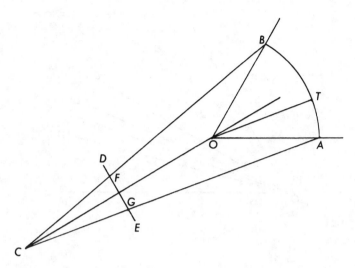

FIGURE 4.60

to 0 at 180°. The construction is equivalent to asserting that

$$\cos\left(\frac{\theta}{3}\right) = \frac{1 - 2D^2(4 - D^2)}{(6 - D^2)^2}$$

where $D = 2\sin(\theta/4)$.

Trisection

By T. M., 1935

In some ways, this is the most incredible trisection in this Budget. M., 21 years old at the time, managed to get his trisection and himself onto the Associated Press wire, and I have three clippings from August and September of 1935 with headlines

CLAIMS ABILITY TO TRISECT ANGLE
DETROIT MATHEMATICIANS STIRRED BY YOUTH'S METHOD

PAGE EINSTEIN! MAN CAN TRISECT ANGLE

YOUNG STUDENT SHOWS HOW HE TRISECTS ANGLE
MATHEMATICIANS STIRRED BY DEMONSTRATION

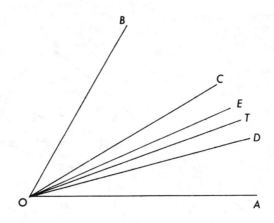

FIGURE 4.61

I have two pictures of M., one with him dressed in jacket and tie, and one in shirtsleeves only, standing in front of a blackboard containing his construction, grinning his head off. There is nothing incredible in that, you may think, it being every American's right to do as much self-promotion as he or she wishes and can get away with. True, but look at what was being promoted! His construction consists of making four bisections which allow an angle equal to five-sixteenths of the original to be constructed. M. then asserts that angle is the trisection. That is, he says that $5/16 = 1/3$. Or, clearing the fractions, that $15 = 16$. That is incredible.

Even though this trisection does not deserve a diagram, it is my duty to produce one. In Figure 4.61, OC bisects angle BOA, OD bisects COA, OE bisects COD, and OT bisects EOD. Angle TOD is thus one-sixteenth of angle BOA; added to angle DOA, which is one-fourth of BOA, we get angle TOA, five-sixteenths of the original angle. The error is huge, exactly one forty-eighth of the angle being trisected.

Trisection

By J. C. M., 1931

A yellowed clipping, taped to a yellowed sheet of paper:

FAYETTEVILLE, Ark., Sept. 15 (AP)—A mathematical feat hitherto considered impossible trisection of an angle with compass and straight edge—was claimed yesterday by a high school instructor, who said he proved it 10 years ago but dropped it because of a "who cares" reaction.

M. produced an eight-page printed pamphlet, with subtitle

An "Irresistible Force"
of geometric demonstration
vs.
The "Immovable Mass"
of a trigonometric equation

It is strange that a high school teacher of mathematics, which is what M. was, would join the crowd of trisectors who deny the validity of trigonometry.

The construction (Figure 4.62) is one in which the temptation to bisect had been resisted. Drop the perpendicular BC, draw the equilateral triangle BCD, make $|BE| = |BC|$, and join E and D to get the trisection point: a simple construction which is good near 60° (an error of only 2′ at 63°) but which is very bad for small angles, not the way that errors usually go.

The clipping concludes

Mathematics instructors at the University of Arkansas withheld comment pending further study and presentation of additional proof.

Shame, oh shame, mathematics instructors at the University of Arkansas! You knew better. Why the timidity, why the cowardice? Surely it is the purpose of all universities, including the University of Arkansas, to pursue and proclaim the truth without failures of nerve. Oh, shame!

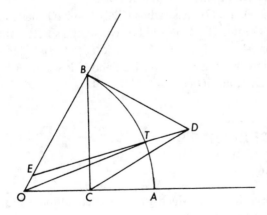

FIGURE 4.62

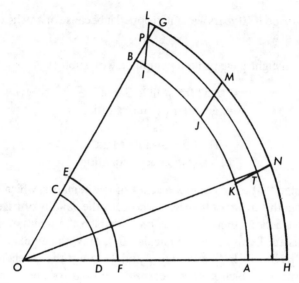

FIGURE 4.63

Untitled Trisection

By N. M., 1981

Here is another of the few trisections in which the angle to be trisected is not bisected first. Its idea is to get a chord that is too short for the trisection, another one that is too long, and then interpolate. It is not a bad idea, the error being less than 20″ for angles up to 81°, but it grows quickly thereafter, reaching 2′ at 90° and a whole degree at 135°.

In Figure 4.63, arc CD has radius $|OA|/3$, arc EF has radius $5|OA|/12$, and arc GH has radius $7|OA|/6$—ad hoc constants, spoiling the organicity of the construction. $|IJ| = |JK| = |KA| = |CD|$; $|LM| = |MN| = |NH| = |EF|$; P is the intersection of IL and BG; and T is where KN intersects the arc with center O, radius $|OP|$.

Trisection of the Angle

By R. A. O'B., 1982

O'B. enclosed his card, which gave his occupation as "Laborer and In-ventor." His method of trisection was to take the angle, cut it out, and roll it into a cone with a circular base. Then trisect the circle, easily done by stepping around the circumference with compasses set to the radius of the circle; this will give you a regular hexagon inscribed in the circle, and taking every other point divides the circular arc into three equal pieces. Then unroll the cone and voila! instant trisection. I complimented O'B. on his originality in getting out of two dimensions to do the trisection, but disappointed him by saying that the scissors were not one of the approved tools for geometry. This idea must be in the air, because in 1985 another person wrote to the Mathematical Association of America giving the same trisection and asking whether anyone else had ever thought of it.

Trisection

By A. G. O., 1985

This trisector would not quit. After being told that his construction was not (and could not be) accurate, he sent me another one. After it was similarly rejected, here came *another.* They were significantly different, and he may be capable of producing yet another. He was as prolific with letter writing as he was with trisections. He has also constructed the regular heptagon, as impossible with straightedge and compass as the trisection.

Here is his nicest construction (Figure 4.64). Bisect angle BOA with OC, draw arc DE with radius $(1/3)|OA|$, make $|OF| = |OD|$, and extend FG to get T. Simple, equivalent to

$$\tan\left(\frac{\theta}{3}\right) = \frac{A\left(1 + \sqrt{8A^2 + 9}\right)}{\sqrt{8A^2 + 9} - A^2} \qquad \text{where} \qquad A = \frac{\sin\theta}{1 + \cos\theta}$$

and never in error more than $10'$ for acute angles.

The Trisection of Any Angle with Straightedge and Compass

By W. O., 1959

O., a medical doctor then aged 85, wrote

Our school system needs an awakening. The shady and false TV and radio programs and ads would cease to pay their cost if teachers gave due emphasis to the art of thinking things thru as well as memorizing.

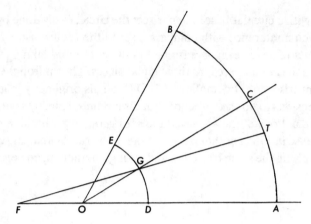

FIGURE 4.64

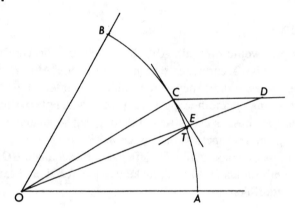

FIGURE 4.65

How true! But the author's conclusion—that his trisection was not accepted because of the closed minds of teachers—did not follow.

In Figure 4.65, OC bisects angle BOA and CD is parallel to OA with $|DI| = (1/2)|OA|$. OD gives the

$$\tan\left(\frac{\theta}{3}\right) = \frac{2\sin(\theta/2)}{1 + 2\cos(\theta/2)}$$

trisection that is so easy to find, but O. did better. Construct a perpendicular to OC at C; it intersects OD at E. A line through E parallel to OC intersects the arc AB at T. The error is only 3′ at 60°, 8′ at 90°, and is never more than 20′ for any angle from 0° to 180°.

Trisection of an Angle by Ruler and Compass

By A. M. P., 1963

In Figure 4.66, OC is the bisector and $|CD|$ is one-third of $|CA|$. Clearly, trisecting the chord will not give a trisection of the arc, so extend CA to E, where $|AE| = 5|CD|$ and locate T on arc AB, $|ED|$ away from E. The error is less than 10′ for acute angles, and less than 20′ for angles up to 180°. I think P. was one of my successes:

> My initial goal was to achieve a construction that accomplished the trisection. Since my method turned out to be a "close approximation over a wide range of angles" [a quote from me] it appears that is what I shall have to accept.

> Attempting to trisect an angle has been an exhilarating experience.

P. sent me a Christmas card for many years, always asking when this book will appear.

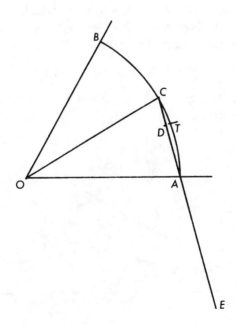

FIGURE 4.66

The [P.] Arc Used in Trisecting an Angle (1957)
Multisecting an Angle (1958, Revised 1959)
The Trisection Problem Solution:
Statement, Analysis, Proof (1963)
Third Supplement to Multisecting an Angle (1963)

By D. P.

P. was a lawyer, evidently successful enough to be able to spend the money needed to have his many booklets printed and distributed. He followed the usual pattern, learning of the trisection as a boy and returning to it after a lapse of 40 years.

> I talked with several mathematicians and they discouraged me. They said it can't be done. I remember that someone had said, "When a person says it can't be done they are merely saying that they can't do it," and the Service Man's Slogan, "The hard things we do immediately, the impossible takes a little longer."

His construction looks very much like the Archimedean trisection. There is no evidence that P. knew about it, but it is possible that he came on it by himself: I have corresponded with one person who reinvented it, and I have another example in my files. In Figure 4.67, start with a 30° angle at C, bisect it, and put D anywhere on the bisector. Determine O, E, and A by making

$$|CD| = |DO| = |OE| = |OA|.$$

Put in angle BOA, the angle to be trisected, and erect the perpendicular OF. Find the center of the circle that passes through F, E, and A; call it G and

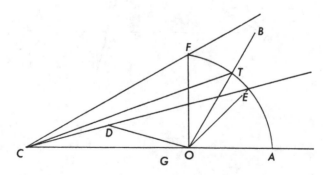

FIGURE 4.67

draw the arc FEA with center G. This intersects BO at T and angle TCA is the trisection.

The approximation is very good for acute angles, with maximum error $5'$. It would be even better if the angle at C were $45°$ instead of $30°$ and better yet (maximum error $2'\,39''$) if the angle were $41°$. But there is no way to construct a $41°$ angle without something like a protractor, and if you had a protractor you would not need the construction.

Beyond the Magic Circle, or, How to Trisect Any General Angle

By R. P., 1973

In this lively booklet, the author tells us how he learned of the problem, read that it was impossible to solve, and then solved it. He did it much more quickly than those trisectors who take years:

> It has taken me 36 days of intense meditation to conquer this ENIGMA of Elementary Geometry. So after more than 2000 years the problem has been solved.

He gives no proofs:

> I must admit I know nothing of Cartesian Coordinates. I have no knowledge of Sines and Cosines; Cosecants or Cotangents. Little does this matter.
>
> The PROOF is in the doing.
>
> It has been done!

The "Magic Circle" of the title

> is only an invisible ring that holds all Mathematicians together in a "groupthink" conglomerate that goes round and round in circles inside the Magic Circle.
>
> Thus, the confusion of tripping over their own formulas and proofs.

In Figure 4.68, OC is the bisector of angle BOA, and $|OD|$ is three times $|OA|$ less the distance from C to the arc BA. T is the intersection of an arc with center O and radius $|OD|$ and one with center D and radius $|AB|$. The error is at most $17'$ for acute angles and $40'$ for angles less than $180°$. The construction

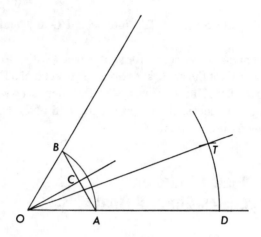

FIGURE 4.68

is equivalent to claiming that

$$\sin\left(\frac{\theta}{6}\right) = \frac{\sin(\theta/2)}{2 + \cos(\theta/2)}.$$

Some Trisetn Properties o Typical Angles in Universal Geometry

By L. S. P., 1982

No, there are no misprints in that title. This is part of volume 14, number 9, of a periodical, *Selfhood Architecture*, written and published by P. His writing style was unusual:

> I also feel that h s 'n college students as well as genl public should b aware o my method 4 trisecting genl angle hence 1st item on practical agenda o study should include preparing acceptable paper on this subject 4 their edific8n.

It is like those advertisements for Speedwriting (where have they gone?)

> If u cn rd ths u cn gt a gd jb n hi pa

though not as extreme. Even though the words are intelligible when read slowly, meaning does not always get through. Here is Step 4 in his "Trisectn Construtn"

> Select 3 = unit O s by series o 'successive approxim8ns' by arriving @ 1 radius taken 3 times from pt o intersection o bisector-subtended

arc—1 leg o angle recognizing according 2 completeness property as it applies 2 plane geometry that a) Every span has = radii multiple spans incldg = radii multiple 3 span b) In any upper = radii span there is shortest upper = radii multiplr span Similarly c) if span has lower = radii multiple span d) Shortest upper = s longest lower = radii spans respively (resp).

Did you get that? P.'s diagrams had too many letters, lines, and circles for me to be able to follow them. I hope that P. has returned to Selfhood Architecture, whatever that is.

[P.]onian Theory of Relativity as Applied to the Science of Mathematics: Showing the Harmonious Co-Ordination of Moving Points in the Complexity of Circles

By R. E. P., 1953

The author of this 24-page booklet wrote

In this booklet is exemplified a new theory, which is a key solution to numerous here-to-fore impossible mathematical problems. The [P.]ONIAN THEORY OF RELATIVITY, as strange and miraculous as it may seem, is so simple that little children should be able to perform with understanding.

There is a trisection of a 30° angle which is perfectly accurate, but whose construction depends on the step

With O as a vertex, construct an equilateral triangle FOG; its base side being perpendicular to OE and subtending the 90 degree arc CED.

The "and" is the trouble; it is impossible to satisfy both conditions with straight-edge and compass. You need the Archimedean straightedge with two scratches on it.

When I wrote to P., his wife wrote back saying that he had died and that she had quite a few copies of his booklet still; since P. had wanted it to have as wide a distribution as possible, she would be happy to send them to me— 2,000 I think there were—for the cost of postage alone. How could I refuse the generous offer? I asked her to send 200, she did, and I still have a few left.

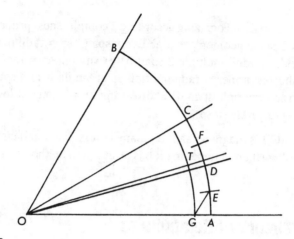

FIGURE 4.69

Trisection

By S. L. R., 1985

The angle trisected, the circle squared, and the cube duplicated, all on five handwritten sheets.

For the trisection, in Figure 4.69, bisect the angle at C, bisect angle COA at D, bisect angle DOA at E, and bisect angle COD at F. Draw EG parallel to BO; draw an arc, center O and radius $|OG|$; and locate T on it so that $|GT| = |EF|$. The error is $9'$ at $45°$ and steadily increases to over $1°$ at $90°$. The construction consistently undertrisects, so here is a challenge to trisectors: move T just a little bit farther along the arc—$(1/3)|AE|$ may be a good distance to try—and see whether you can get it right on.

Circular Arcs Which Seem to Trisect Angles

By L. R., 1981

This was submitted to a national mathematics journal. Why some trisectors do not bother to consult local authorities is not clear; here the diffidence of the title may reflect the diffidence of the author. It is easier to get the bad news by mail.

If only (Figure 4.70) you could find a circle, center at C (on the bisector of angle BOA) such that the distance from D to E is the same as the distance

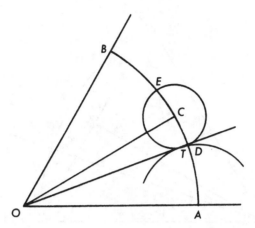

FIGURE 4.70

from A to D, then the tangent OT would indeed be a trisector. The reason is that if angle COT is θ, then angle TOA is 2θ; angle COA is 3θ, the whole angle is 6θ, and the trisection is done. L. tried to sneak up on the desired circle by a process of approximation, but the circular arcs of his title were not quite circular.

Problem: To Trisect an Angle by (1) Carpenter's Square Alone (2) Auxiliary Curve (3) Straight Edge and Compasses

By M. J. R., 1928

(1) and (2) are perfectly possible. On (3),

Of these three (3) has been demonstrated to be impossible by a gentleman in the middle of the nineteenth century. Since then few attempts have been made at its solution. . . . Certain new data have been found which I shall not exhibit fully in this sketch but shall enlarge upon later in another cover.

I have not seen the other cover, so it may be unfair to include the author in this Budget; he may have had second and better thoughts. The great mathematician Legendre once thought that he had proved Euclid's parallel postulate (another impossibility, but not proved to be impossible until after Legendre's

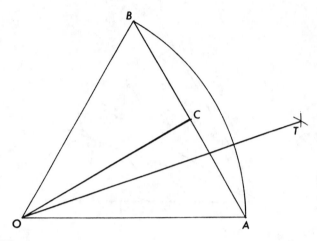

FIGURE 4.71

time); just before he was to have presented his proof in public he is reputed to have said, "I will have to think further." He did, and never presented his proof. If I am haunted by the ghost of an angry R., I will deserve it.

Untitled Trisection

By M. F. R., 1973

In this simple trisection (Figure 4.71), OC bisects angle BOA and T is determined by making $|CT| = |AT| = |AC|$. The point found is the same as in the trisection of C. H., an unusual coincidence. R. gives a 30-step proof in statement-reason form which makes the error (it is at step 23) easy to find.

The Trisection of an Angle—Triple-arc Method

By J. C. R., Ph. D., 1971

R. did not specify the field of his Ph. D. degree.

The construction (Figure 4.72) is simple: $|OA| = |AC| = |CE|$, OG bisects angle BOA, and T is found by drawing an arc with center E and radius

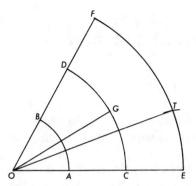

FIGURE 4.72

$|CG|$. It is equivalent to saying that

$$\cos\left(\frac{\theta}{3}\right) = 1 - \frac{8}{9}\sin^2\left(\frac{\theta}{4}\right).$$

The construction is fairly accurate for small angles, the error being 7′ at 60°, but it is more than 26′ at 90°, and it increases rapidly thereafter.

Later, a professor of mathematics at the University of [C.] told me that R.'s Ph. D. degree was in mathematics! And he made his living teaching mathematics! The professor wrote

> He was routed to me and I spent several sessions with him. His error was easy enough to spot but that was not at all convincing, nor did it bother him at all when I applied his construction to 60° and showed that you get 19° 46.8′. (He said, "applying numbers to what is pure geometry is inappropriate.") He kept bugging me and (worse) bugging the department chairman so much that finally we arranged for him to give a departmental colloquium. The place was jammed with mostly students who had come to see the fun. They gave him a perfectly dreadful time but it didn't faze him a bit—in fact he gloried in the whole thing. (I'm sure that on his vita now appears: Lectured for the University of [C.] Colloquium.) For many months thereafter we were bombarded—more trisections, circle quadratures, regular nonagons, etc. .... We have heard nothing from him for a long time.

Yes, trisectors and other cranks are often eager to have audiences. However, as was shown by the University of [C.], there is nothing to be gained, by anyone, by giving them one. It is, in fact, an act of cruelty.

The Trisection of an Angle
with Compass and Straight Edge

By H. R., 1983

R. submitted his manuscript to a vanity press, which I hope declined to print it. He had evidently shown it around, because he wrote

> I will prove that statement by trisecting a 60° and a 75° angle, with just a compass and straight edge, and proving the constructions by Plane Geometry. They cannot be proven by trigonometry. The only problem that can be proven either by geometry or trigonometry is the 45° angle trisection.

Someone had no doubt used trigonometry to find, as I did, that his trisection of a 60° angle produced one of 20° 2′ 15.6″. So, like many other trisectors, he pronounced trigonometry illegal. Do these trisectors think that because they use only straightedge and compass, sin 30° is not 1/2, or 2 + 3 is no longer 5? It is a mystery to me.

His construction is too complicated to explain, but here (Figure 4.73) is an unsimplified picture of it.

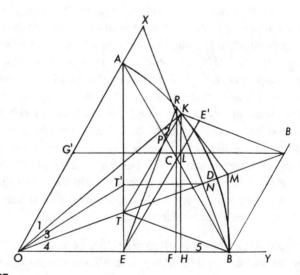

FIGURE 4.73

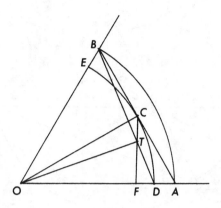

FIGURE 4.74

Trisection of an Angle, Using Straight Edge and Compass, Only

By S. G. R., 1975

In Figure 4.74, OC bisects angle BOA, and D and E are on an arc with center O and radius $|OC|$. Drop a perpendicular to OA from C, draw BD, and their intersection is T. This is yet another version of the Steinhaus-and-many-others trisection.

Doubling the Arc and Trisection of the Angle

By W. H. R., 1904

In the introduction to this four-page pamphlet, W. H. R. made the typical statement

In giving out these solutions, I wish to say that I have asked the "advice or consent of no one." The solution for the trisection of the angle by Euclidean geometry has been an unsolved problem since the time of the great mathematician Euclid himself, and I am conscious that I am assuming much when I claim that I have found a solution for it.

The construction (Figure 4.75) is fairly elaborate. Draw the circle ABC and make DOE bisect angle BOA. F is the midpoint of OC, and G is found by drawing an arc with center F and radius $|OA|$. The large arc has center G and radius $|DG|$; it determines H and I. Finally, OT is drawn parallel to IH. Elaborate constructions, and those that go behind the angle, tend to be more

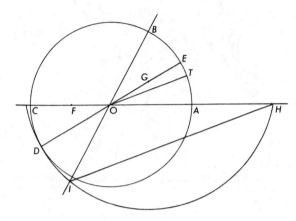

FIGURE 4.75

than usually accurate, and this is one of them. The error is only 2′ 20″ at 60°, and the maximum error for all angles for 0° to 180° is a mere 4′ 33″.

Untitled Trisection

By A. B. R., 1959

This (Figure 4.76) is another elaborate construction. Given angle BOA to be trisected, draw circle $ABCD$. Bisect angle BOA with OE and draw OF perpendicular to OE. G is located on OE extended by making $|AG| = |OA|$. Draw FG. (B will not lie on FG in general.) H is found by making $|GH| = |OG|$ and the arc through I and J has center H and radius $|OA|$. J is found by making $|IJ| = |BE|$, and joining J and C locates K on OI. Finally, the intersection of the circle with KD gives the trisection point T.

For all that work, the construction is not terribly accurate. The error is as large as 16′ for acute angles.

Trisection

By E. R., 1947

This is an eight-page pamphlet, printed in German, English, and French, accompanied by four beautifully drawn diagrams. R., "officier d'Académie Française," has also squared the circle and duplicated the cube.

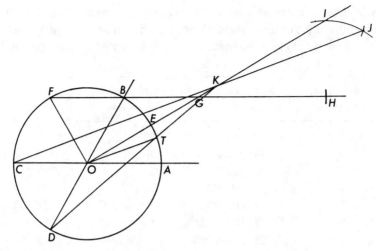

FIGURE 4.76

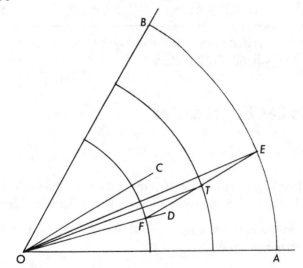

FIGURE 4.77

His construction (Figure 4.77) starts by dividing OA into four equal parts and drawing arcs with center O. OC bisects angle BOA, OD bisects angle COA, and OE bisects angle COD. T is the intersection of EF with the third arc.

Since angle DOA is one-fourth of angle BOA and angle EOA is three-eighths of angle BOA, one-third of the angle must lie in between, so it is not surprising that the construction is accurate. It could have been made more accurate by making one more bisection and getting T between five-sixteenths and three-eighths of the original angle. Of course, with sufficiently many bisections it is possible to get as close to a trisection as you want to: for a 60° angle we have

Number of bisections	Closest fraction to 1/3	Best approximate trisection	Error
1	$1/2 = 0.5$	30°	10°
2	$1/4 = 0.25$	15°	5°
3	$3/8 = 0.375$	22° 30′	2°30′
4	$5/16 = 0.3125$	18° 45′	1°15′
5	$11/32 = 0.34375$	20° 37′ 30″	37′ 30″
6	$21/64 = 0.328125$	19° 41′ 15″	18′ 45″
7	$43/128 = 0.335938$	20° 9′ 23′	9′ 23′
8	$85/256 = 0.332031$	19° 55′ 19″	4′ 41′
9	$171/512 = 0.333984$	20° 2′ 21″	2′ 21″

and so on: each bisection halves the error. The error in the R. construction is 1′ 14″ at 60° and still only 32′ 28″ at 180°.

The Angle Can Be Tri-sected

By H. F. S., 1973

This author discovered a trisection that uses more than straightedge and compass, and he did not delude himself by thinking otherwise. He wrote

> To date, and to my knowledge, there has not been a solution for the tri-section of any angle.

He was wrong about that, but he was right that his construction does exactly trisect angles. However, as he said,

> a prime requisite [is] the ability to use the Pencil, Ruler, Compass, and French Curve.

That last tool is not part of Euclidean geometry.

S. was the subject of an article in *Newsday* in 1974:

> [S.], who is 64, dropped out of junior high school in Brooklyn when he was 15 to help support his family.

He took some drafting courses at a Brooklyn YMCA and attended a technical school, where he learned to work with sheetmetal. He became a jack-of-all-trades in the electrical sign industry, sketching the letters, laying out the patterns, supervising the installation. "That's where I became interested in math," [S.] says. Four years ago ... he took up the trisection problem in earnest, often spending at least 12 hours a day on it.

"I didn't even know who Euclid was when I started," he says. "I found out how ignorant I was in arithmetic and algebra, so I taught myself to read a slide rule." He sent copies of his first trisection work to virtually every state university in the country. Only three replied.

That is no wonder, because his construction is complicated, cluttered with lines, unappealing to the eye, and hard to follow: it took me quite a bit of effort to discover that S. had indeed trisected by using an auxiliary curve. S. is the trisector who was granted U. S. patent 3,906,638 for his trisection.

I imagine S.'s trisection history is similar to that of many other trisectors, but it is to his credit that he did not convince himself that he had found a straightedge and compass trisection. It is a problem what to say to trisectors like S.: tell them that their work is correct, but not new; congratulate them on their rediscovery and say that it is of no interest? Not much reward for four years' work! If only there were universal mathematical literacy, the problem would not arise and the years of time would not be wasted. I am afraid that universal mathematical literacy will be achieved at about the same time as inexpensive antigravity and commercial teleportation.

[S.] Theorem—Any Angle Can Be Trisected

By K. B. S., 1972

Here is a chance for you to do trisection detective work. I think there must have been another sheet explaining the construction of Figure 4.78, but I do not have it; the diagram is all there is to go on. Angle ABC has been approximately trisected, and with straightedge and compass alone, but how the construction was done would have to be deduced from the diagram. S. had at least lettered his points in order, I think, so that provides valuable information. If you succeed in reconstructing the construction, I would be pleased to know how it goes.

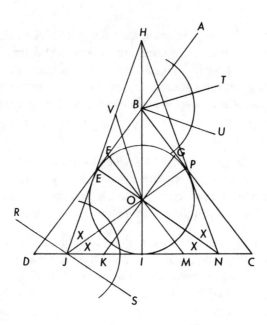

FIGURE 4.78

To Tri-sect an Angle
Using Compass and Ruler Only

By F. K. T., Esq., 1980

T. was neither a lawyer nor a squire; the *Esq.* is only a rhetorical flourish. The last step of the trisection was "Resulting isosceles triangle trisects 360° angle into equal arcs of 120° each." What he had done, and it was all that he had done, was to construct a 120° angle. The same thing can be done more quickly by making an equilateral triangle and extending one of its sides. The amount of ignorance in the world is commonly underestimated.

The Trisection of an Angle

By J. T., 1914

This is a 12-page pamphlet, nicely printed and written in good theorem-proof form. T. confined himself wholly to his construction and his attempted proof of its correctness. The construction is confined to circle *ABC* (Figure

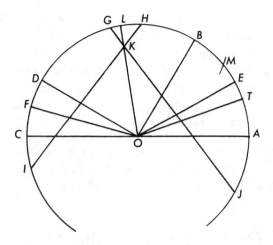

FIGURE 4.79

4.79). Erect a perpendicular to OB at O, intersecting the circle at D. OE bisects angle BOA and OF bisects angle DOC. G is 60° clockwise on the circle from F (gotten by marking off an arc with radius $|OA|$ and center F) and H is 60° counterclockwise from E. I is determined by making $|CF| = |CI|$ and J by making $|AJ| = |AE|$. (Regardless of the size of angle BOA, the inclination of HI is 37.5° and the inclination of GJ is $-37.5°$.) The intersection of GJ and HI determines K, and L is on OK extended. M is 60° clockwise from L, and T is found by bisecting angle MOA.

The construction is unusual in having its largest errors for angles near 0° and 90°, but the error is no more than 7' for angles between 20° and 70°.

On Geometrical Trisection of Angles: A Problem from Antiquity Renowned as Insolvable

By L. J. T., 1934

I pride myself on being able to subdue any trisection, but this one conquered me. I attacked it two or three times, but gave up each time; my pride is kept intact by my thinking that if I *really* made an effort, I could get it. T. wrote, after "the angle is geometrically trisected"

In passing, it may be said that considerable proof, geometrical by design and elusive by nature, seems to avail a natural substantiation about those controversial points, requiring the presence of a magician to determine by the hat and its contents, but constituting the principal basis for the fundamental idea upon which some previous work in this direction on the part of the writer has been tentatively predicated in prankish surmises with some modicum of levity.

I was unable to follow that either; perhaps the author wrote about his construction in analogous style. If anyone wants a copy of this two-page work, I will be glad to send it. The copyright has probably expired.

Geometry: The Trisection of the Angle and Theorems and Corollaries Leading to It

By F. T., 1933

This 16-page pamphlet is by the trisector who succeeded in getting a testimonial from a professor of mathematics:

> I have carefully checked your work on "the trisection of the angle" and was not able to detect any fallacy in your work.

Perhaps the professor used this as the quickest way of shutting off a correspondence he did not want to begin. Perhaps he was being ironic, the quote meaning that he found no fallacy because he devoted no time to it. I am afraid, though, that the professor indeed could find no flaw and in fact was not aware that the trisection was impossible. Professors as well as trisections can have flaws.

T. was under the usual misapprehension about what *impossible* means in mathematics:

> Therefore, it is apparent that even we, of this twentieth century, have conceded and are convinced that what our ancestors have failed to accomplish in mathematics is an impossibility.

The construction (Figure 4.80) is easy to make: OC bisects angle BOA, D is on OC such that $|CD| = |CA|$, and T is at a distance $|BD|$ from B and D. It locates the same point as the construction made in 1847 by H. L. and found often since then.

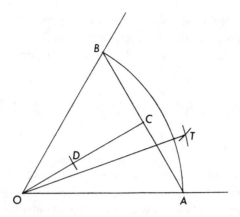

FIGURE **4.80**

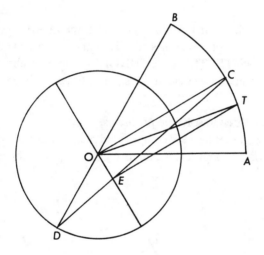

FIGURE **4.81**

The Methods of Trisecting Angles

By T. G. T., No Date

The proof that this trisection is correct is easy to follow, making the error easy to see. The direction in the proof "Now from L, draw LN parallel to VOM, meeting the circle at N" cannot be carried out because the line will not meet the circle at N. Trying to make a line do more than it is capable of is a common mistake of trisectors.

In Figure 4.81, OC bisects angle BOA as always, and the circle has radius one-half of $|OA|$. D is the intersection of the circle and BO extended, and E

is the intersection of CD and the diameter of the circle that is perpendicular to OC. A line through E parallel to OC intersects the circle at the trisection point T. It locates the same point as the J. J. construction.

Method of Construction of Geometrical Structures

By I. T., 1982

The angle trisected, the circle squared ($\pi = 3.146853\ldots$), and the cube duplicated, all by a resident of Greece, where the problems first arose. The drawings in this two-page booklet were complicated but well done, not surprising since

> I am not a Writer nor a Mathematician. I am an Artist. I attended Painting and Sculpture classes for six years at the ROYAL MELBOURNE INSTITUTE OF TECHNOLOGY in Australia. While trying to draw a line in a circle and a square, I saw that some relations exist ...

and he solved the three unsolvable problems.

I was surprised that his very complicated diagram turned out to contain a simple construction: in Figure 4.82, angle COD is one-fourth angle BOA, and

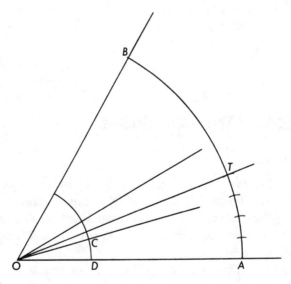

FIGURE 4.82

$|OD| = |OA|/3$. T is located by laying off $|CD|$ four times on the arc AB. It is not bad, as simple constructions go, with error at most $11'$ for angles less than $90°$, and it is equivalent to asserting what students of trigonometry wrongly have asserted, are asserting, and always will assert, namely

$$\sin\theta = 3\sin\left(\frac{\theta}{3}\right).$$

Too bad that neither trigonometry nor trisection is that simple!

I did not get a reply to my letter saying that his trisection was not perfect for a long time, and I thought that my impressive computer printout had convinced him of his error. Not so: eventually he sent me another copy. And then, after another few months, yet another copy (with a few small changes), this one having the marks of being one of a mass mailing. Oh dear, had my reply only stirred him up? Was Greek pride going to refuse to admit error? Since T. sent his material by registered mail, I had to make a trip to the post office to get each of his letters, which can get tiresome. On the other hand, each communication cost T. about $3 in postage. I knew he would weaken eventually, and he did.

Trisection

By L. W. T., 1931

School Science and Mathematics saw fit to publish "An analysis of a purported trisection of an angle with ruler and compass" by H. C. Shepler (43 (1943), 465–467) showing that T.'s trisection, which appeared in the Dubuque (Iowa) *Telegraph-Herald*, was incorrect. The 1931 date of the trisection makes me think that it was inspired by Father C. What mischief he did!

In Figure 4.83, angle BOA is bisected, and OC, perpendicular to the bisector, is constructed. CA is drawn parallel to the bisector, and D is on CA at a distance of $2|OA|$ from A. D locates the point of the trisection that amounts to saying that

$$\tan\left(\frac{\theta}{3}\right) = \frac{2\sin(\theta/2)}{1 + 2\cos(\theta/2)},$$

which has been found so many times, and which can be improved on, as T. proceeded to do. Construct E at a distance $|OA|$ from C, get F at a distance $|OD|$ from O, and draw an arc with center F and radius $|BE|$ to locate T on arc FD. This construction asserts that

$$\sin\left(\frac{\theta}{6}\right) = \frac{\sqrt{5 - 2\cos(\theta/2) - 3\cos^2(\theta/2)}}{2(5 + 4\cos(\theta/2))}$$

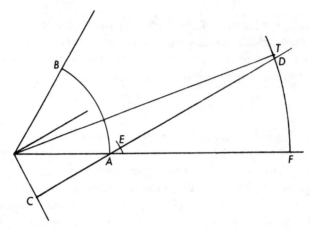

FIGURE 4.83

and is not bad: the error at 60° is a mere 1′ 45″, at 90° only 5′ 12″, and the maximum error for all angles between 0° and 180° is less than 20′.

Untitled Trisection

By L. T., 1974

This is a letter to a department of mathematics, starting

Enclosed is a method, which, I believe, makes it possible to trisect an angle.

The construction (Figure 4.84) is simple: OC bisects angle BOA, and CD bisects the right angle at C. T is located so as to make C, D, and T the vertices of an equilateral triangle. Trigonometrically, the construction says

$$\tan\left(\frac{\theta}{3}\right) = \frac{\sin(\theta/2)\big(\cos(\theta/2) - \sin(\theta/2)\big) + \sqrt{2}\big(\sin(\theta/2 - \pi/12)\big)}{\cos(\theta/2)\big(\cos(\theta/2) - \sin(\theta/2)\big) + \sqrt{2}\big(\cos(\theta/2 - \pi/12)\big)},$$

and for acute angles the maximum error is only slightly more than 8′.

A curious coincidence is that this trisection locates the same point as that of T. S. T., found later in this Budget, though with a different construction. L. T. and T. S. T. were both trisecting at the same time, in the same place, and found the same point, an unlikely coincidence. T. S. T. was an indefatigable correspondent with anyone who would correspond with him, and I think that

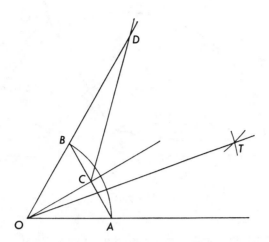

FIGURE 4.84

he had worn out whatever welcome he had anywhere. I conjecture that T. S. T. showed young L. T. his trisection, and then, with or without T. S. T.'s knowledge, L. T. started to circulate his version to see whether he would get any different reception from T. S. T.'s. My letter to L. T. concluded

> By the way, your construction gives exactly the same results as one by Mr. [T. S. T.], who lives in [your city]. A remarkable coincidence!

> There was no reply.

Untitled Trisection

By B. T., 1982

This trisector was old (typical), made complicated drawings that are not pretty (typical), and had been more than half-persuaded by years of correspondence with a patient mathematician that he had found only an approximation (decidedly untypical).

In Figure 4.85, D is the midpoint of OA, E is one-twelfth of the way from O to C, and the inner semicircle has center E, radius $|ED|$. F is the intersection of the bisector of angle BOA with the semicircle, and CF intersects the outer semicircle at the trisection point T. T. is, perhaps unconsciously, trying to locate the Archimedean trisection point. The error is less than $9'\ 11''$ for acute angles, and it is less than $20'$ for angles from $0°$ to $180°$.

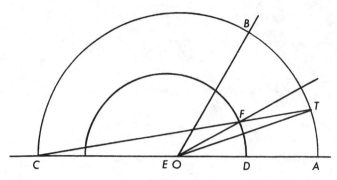

FIGURE 4.85

The construction is simple and accurate, but aesthetically marred. Point *E* is ad hoc—why one-twelfth?—the construction does not flow, it is not organic, and its texture is too nubbly. No one can disagree with that artistic assessment since I am preeminent in the field of trisection connoisseurship. It is easy to be the world's leading expert when there is only one.

Adventures in Geometry

By C. D. T., B. S. E. E., 1974

This is a 173-page hardbound book, with title and author's name stamped on the cover in gold. It contains a trisection, multisections, and an attempted quadrature of the circle. Many examples and proofs are given, and the whole represents what must have been a large investment of time, work, and energy, not to mention money. I wrote to the author, giving him the bad news that his trisection did not and could not work, and I was astonished by his reply, which included

> I read your verdict with mixed feelings, as you can understand. Every-
> one that has "dabbled" in the subject, like myself, hopes he has the
> "real" thing. So I was somewhat disappointed at not being perfect.
> But then on the other hand everyone, as well as every text I read on
> the subject, tried to tell me the impossibility of what I set out to do. I
> consider the time and study I put into it as a rewarding pastime and
> experience. ... I am going to put the whole matter to rest now and
> am satisfied at having obtained instead of a true 20, a close approxi-
> mation of 20.000436 degrees.

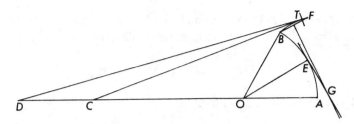

FIGURE 4.86

What an amazing display of sanity!

It is to the credit of his sanity and Bachelor of Science in Electrical Engineering that his construction is one of the most accurate in this Budget. It is unlike all of the other trisections, so it took an original and creative mind to hit on it. In Figure 4.86, OA is extended so that $|OC|$ is twice $|OA|$ and $|OD|$ is three times $|OA|$. OE bisects angle BOA, and the other line at E is perpendicular to OE. F is determined so that $|DF| = |DA|$. G is the point where the perpendicular bisector of BF meets the perpendicular from E. T is the intersection of two arcs: one with center C and radius $|CA|$ and the other with center G and radius $|GB|$. Angle TCA is the trisection. The construction would be hard to draw accurately—G is at the intersection of two nearly parallel lines—but if it could be done, it is accurate indeed. The error is 2″ (that's 2″, not 2′) at 60°, only 12″ at 90°, and still less than 7′ at 180°. This was no accident, but the result of engineering reasoning:

> My method of Trisection is based on the fact that a wheel or circle of radius $3r$ rotating without slip and in circumferential contact with a wheel or circle of radius r will sweep out one third the arc of the latter wheel or circle.

Too bad the trisection cannot be done: if it were possible, T. might have done it.

Untitled Trisection

By J. A. T., 1973

T. was an enthusiastic amateur of geometry, though he knew no trigonometry or higher mathematics. He was more or less convinced that his construction was not exact, but I think he still had a faint hope in the back of his mind that he had in fact found the trisection. He wrote a two-page document,

THE TRISECTION GAME

with 18 paragraphs, starting with

> Only a compass and an unmarked straight edge may be used.

and including

> If you have an error in a basic rule, complication in construction will not eliminate the error. So if you have a very complicated construction pull out the essential lines and put them on another drawing. Any error will probably show up.

> Whatever you do, do NOT waste your cash on copyright, patent, or printer. Postage will be plenty. Maybe you will never get heard. But save your coin or spend it on roses for the soul or something.

> Of course you are crazy. Who isn't? So sneer right back, but sort of smile. It helps.

> Of course you are a darn nuisance. Nobody wants to hear you, right or wrong. They are full, right to the top of their bald, or something else, head. Remember the only reason why we have no trisector's club is because you wouldn't look at anybody else's stuff, either. But go right ahead. Most of the world's advancement in math comes from people who failed to trisect but got sidetracked by what they found. You do something like that and you may be famous. Most likely you will not, but it is the one in the million that did find something that counts, and the world is better.

Good advice, mostly. His reason why there is no American Trisection Association with dues, officers, publications, and an annual convention with papers and parties is right on the nose, but I think he is a bit off about the source of mathematical advancement. There is no record that Gauss or Newton ever attempted the trisection.

By the way, I hope that no one reading this is guilty of that uncouth abbreviation, math. When I hear it, I am tempted to ask the utterer why the subject of grass-cutting has come up. A math is a mowing (whence aftermath)—you can look it up. Why does the ancient and honorable subject of mathematics get its tail cut off? After all, people do not talk about phys, or hist, or Eng. Yes, I

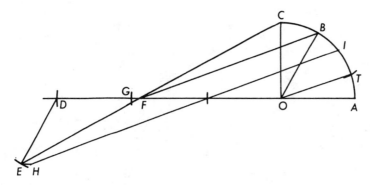

know that mathematics is all of four syllables long, but people educated enough to talk about the subject should be able to handle words of even greater length. However, a language belongs to its users, and if they are going to say "Hopefully, math will be my media" the rest of us will have to go along or risk seeing the pitying smile bestowed on those quaintly and hopelessly out of date.

In Figure 4.87, angle BOC is constructed so as to be one-half of angle BOA and $|OD| = 3|OA|$. DE is parallel to and the same length as BO. $|OG| = 2|OA|$, and H is found by drawing EH parallel to and the same length as GF. HI is parallel to BF, and T is found by making $|IT| = |BI|$. The error is very small for small angles, less than $1'$ for angles up to $58°$, but it increases rapidly, exceeding $1°$ at $90°$.

A Theory of Angle Division

By R. T., 1943

Here is the second and last female trisector in this Budget. She may not deserve the title of trisector, because before plunging into her trisection, she wrote only

I present this theory of angle division as a challenge to all proofs of the impossibility of trisecting an arbitrary angle with rule and compass.

You see, she could always say that she hadn't done a trisection, but only issued a challenge.

The construction is unusual because it allows choice: after drawing the arc BA (Figure 4.88), C can be picked anywhere on the arc as long as arc BC is

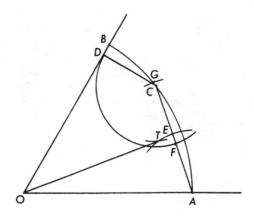

FIGURE 4.88

less than arc AC. I took C to be about one-third of the way along; it is possible that another choice would have made for a more accurate construction. After C is picked, join it to A and drop a perpendicular to OB at D. Draw an arc with center C and radius $|CD|$ and another with the same radius and center at A. The arcs determine points E and F. Trisect the segment EF and draw an arc, center A and radius $|AC| + (1/3)|EF|$, cutting arc BA at G. Finally, draw an arc with center G and radius $|CD|$, cutting E's arc at T, the trisection point. The error increases steadily as angle BOA increases; it is 10′ at 60° and 36′ at 90°.

A General Method for the Geometric Trisection of Angles and Arcs

By W. H. T., 1902

Even though T. styled himself "Private Tutor in Mathematics," he was unaware that the trisection is impossible:

> Those who are skeptical should offer more than rhetoric and argument in order to disprove geometric facts.

That sounds as if he had been told of the impossibility. The following sounds as if someone had showed him that he was wrong by using trigonometry:

> If the points of trisection . . . are not respectively equally distant from the sides of two-thirds of the given angle, they should show by ruler

and compasses that these points respectively are unequally distant from said sides.

Ruler and compasses only—no fair using trigonometry!

The construction is exactly the same as the one of F. T.: same method, same diagram, same point determined. It is in the highest degree improbable that F. T. knew of W. H. T.; the duplication must be one of those coincidences that happen every so often. Even an unlikely event, given long enough to occur, must happen.

Untitled Trisection

By B. H. T., 1974

The author of this construction had written a 40-page book, all by hand, giving the construction and a proof of its correctness. The sheets in the book are in several colors, with illuminated capitals and decorative scrolls; they are works of art and must have taken hours of labor to produce.

T. had circulated his work to many mathematicians. He wrote

> My original—and continuous—purpose was to find a solution for trisecting the angle using only the Euclidean precepts. I have been reminded many times by well-wishers and not-so-well wishers that such a course is impossible. As you can imagine, there were divers and multitudinous trials since my experimenting began in the fall of 1929, so that, not only were a legion of anticipated possibilities struggled through, but at length, after a process of elimination a pattern eventually began to take shape. This was in the late 1940s.

> I have been reminded more than once of the pronouncement made by Wantzel in 1837. It seems that the edict passed upon trisection by this erudite gentleman had the unfortunate effect of closing the minds of many otherwise brilliant men who have been discouraged by it.

He also wrote

> I became engrossed with it [the trisection] to the exclusion of the other problems [the quadrature of the circle and the duplication of the cube], and to practically all else. I doubt if any one person has spent more spare time with the problem than I.

Over the course of more than 40 years, T. had no doubt become so obsessed with the trisection that nothing that anyone could say to him would make any impression. As evidence of that, he later wrote to another person

So again I contacted mathematicians, and, this last year, sent copies of a new book of drawings to two math professors, one in Indiana and one in Southern California. The Indiana professor computerized the problem, but, according to his returns, had fed incorrect data into the computer.

I have come to expect advice from mathematicians—that my effort has been nothing more than a waste of time—as standard procedure, along with admonitions against further attention to it, as "solution of the problem was proven impossible by Wantzel in 1837." I have never been able to find such a proof. However, their advice never fails to discourage me immensely, and I find myself on the verge of giving up the whole thing and conceding that they must be right.

But the setbacks are usually of short duration. After a few days I find myself back at it again, reviewing the delineation step by step. This is usually brought on by the spark of a new idea, or the search for a weak spot. As a result of the latter, a better way of expressing a statement or reason often evolves. But as always, a thorough review of the problem's description usually results in a renewal of my belief in it.

Some people have minds that cannot be changed, for any reason. Death is the only thing that would kill T.'s belief in his trisection.

T.'s construction, like R. T.'s, contains a variable. In Figure 4.89, arcs BC, CD, and DE have the same length, which can be anything greater than one-third of the arc BA. F is on OA extended at a distance $|BE|$ from B. G is

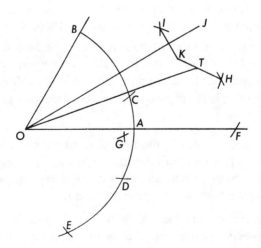

FIGURE 4.89

the intersection of two arcs with radius $|OA|$, one with center B, the other with center F. H is also the intersection of two arcs: one with center G and radius $|OA|$ and the other with center B and radius $|BD|$. I is, too: center G, radius $|BG|$ and center B, radius $|BC|$. Drop a perpendicular from I to OJ, the bisector of angle BOA, and extend it an equal distance on the other side to get K. The trisection point is then the midpoint of segment KH.

If arc BC is exactly two-thirds as long as arc BA, then the trisection is exact, but of course if you could construct arcs two-thirds as long as given arcs, you would have solved the trisection problem. The closer you get to two-thirds, the better the approximation is: I took the length of BC to be 65% of the length of BA. The error is then only 1′ 12″ at 60° and 4′ 11″ at 90°, though it grows rapidly after that. If BC is 60% as long as BA, the error is almost 14′ at 90°, so the construction is quite sensitive to the length of BC.

Trisection

By T. S. T., No Date

This author applied his method only to the 60° angle, but it can be applied generally. His diagrams contain more lines than those of any other trisector I have come across. I was pleased to be able to understand, after a while, what was going on in them.

His construction (Figure 4.90) is to draw OC, the bisector of angle BOA, and find D on OC so that $|CD| = |CA|$. T is then found by making B, D, and T the vertices of an equilateral triangle.

It was only after I had calculated the errors for various angles that I noticed that they were exactly the same as the errors in the trisection by L. T. It was then that the following comment became significant:

> I am choosing to wait for a meeting with the young "genius" at Cal Tech, before I release any more diagrams to anyone. This is the ONLY person who has offered to meet with me in person to discuss my theories. Others only REPEAT what they have been taught, "IT CAN'T BE DONE, SEE SO AND SO."

Both L. T. and T. S. T. had constructions involving equilateral triangles; it is hard to resist the conclusion that L. T. was "genius" enough to recognize that the construction could be used on angles other than 60° and that he modified it slightly. T. S. T. was convinced that the trisection is connected with tarot cards, the Kabbala, and other mystical ideas. Trigonometrical disproofs of his trisection were met with

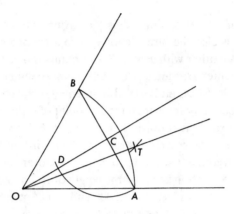

FIGURE 4.90

Trigonometry is a man-made concept. This does not mean it will not function, but it lacks the ability to perform like the natural law. In my demonstration I am using natural law.

His attitude is shared by some other trisectors, who seem to think that Euclidean geometry describes reality and is therefore natural and right. Why they do not think that trigonometry shares the same sacred quality—it describes reality just as well as Euclidean geometry—is not clear, unless it is because they learned about geometry in school when they were young, but never encountered trigonometry.

Untitled Trisection

By S. T., 1981

Telephone calls from trisectors are always unwelcome—it is much easier to give them the bad news in writing—and telephone calls from trisectors who are only 150 miles away and badly want to come and visit in person are ordeals. S. T. was good enough not to insist (Hoosiers are better people than the average), and after being told that his construction was not accurate he said that he was convinced that he found only an approximation. It was the computer printout that did it, I think: computers still inspire some undeserved awe. I count T. as one of my successes. One of my few: I can still count the number with my fingers, and I suspect that it will be a long time before I need to take off my shoes.

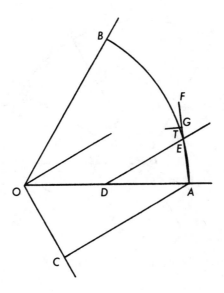

FIGURE 4.91

The construction (Figure 4.91) starts with OC, perpendicular to the bisector of angle BOA. CA is parallel to the bisector, and D is determined by being equidistant from O and C. It is only because BOA measures 60° that D lies on OA. A line through D parallel to the bisector intersects the arc at E, and F is on the line through A and E with $|AF| = |OC|$. G is found by trisecting AF. G is a good enough trisection point that many trisectors would have been satisfied with it, but the author improved it by using T, where the perpendicular erected at G intersects the arc. The maximum error for angles between 0° and 180° is 3′ 24″, and a 60° angle is trisected with an error of only 1′ 4″.

Trisection General de L'angle; La Trisection Vaincue

By G. T., 1973

T. concluded that his trisection provided

la preuve mathematique de l'existence de DIEU.

Mathematical proofs of the existence of God are not as common as they once were. The reason may be that the development of non-Euclidean geometry convinced everyone that mathematics had no direct connection with nonmathematical reality. Or, it may be that fashions have changed. T.'s proof is all mixed

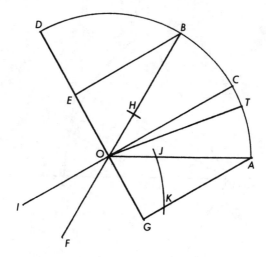

FIGURE 4.92

in with references to the Kabbala and can be fully understood only mystically. His idea seems to be something like "all those mathematicians from Pythagoras on have been mistaken about or unable to do the trisection; I have done it; it has been revealed to me by God; hence God exists." But that is much more logically put than T. put it, and he might denounce my exposition as distorting terribly his proof.

In Figure 4.92, OC bisects angle BOA, angle DOB equals angle BOA, and BE is parallel to OC. F is on BO extended, with $|EF| = |OA|$. AG is parallel to OC. H is on an arc with center F and radius OA, and I is on OC extended with $|HI| = |OA|$. This trisection was on the point of being one of unparalleled virtuosity in which the compass never had to be reset, but T. was unable to carry it off: the setting has to be changed to $|JK|$ to locate T, which is that distance from A. The error is less than 6′ for angles up to 60°, but the accuracy deteriorates quickly as the size of the angle increases.

Trisection

By C. T., 1972

A mathematician in the Netherlands wrote to me

In February 1972 we went to a Science Fair in Taipei. At one of the stands was a trisector. How he got there I forgot. He was ignored

by the jury. He was also very angry that they judged his work without seriously studying it. He threatened to appeal to the Ministry of
Education. This man, I can still see him. Rather heavy built, grayish,
fifty or more probably over sixty, not very tall. He was an elementary school teacher and had been working on the problem for twelve
years. When told that it was impossible on algebraic grounds, he kept
saying that he knew it was *algebraically* impossible, but that did not
imply that geometric methods would not work. In fact, he said, his
work showed that geometric methods would work. My wife, young
and energetic, spent the rest of the day and part of the evening going
with him through his proof. Finally there were three points that had
been assumed collinear without proof.

East is East and West is West, but the trisection can transcend that. The preceding could have been written about someone in Missouri, it is so typical of
trisectors and trisections. One cultural difference shows, however: the author's
diagram was such a gorgeous example of calligraphy that I had it framed.

In Figure 4.93, angle AOC is the same size as the angle to be trisected,
D is on OC extended, OE is perpendicular to OA, and F is on OE. G is on
OA extended with $|DG| = 2|OA|$ and $|GH| = |DF|$. The intersection of DH
with the circle locates T. Another accurate construction-behind-the-angle, with
maximum error a mere 1′ 30″ up to 60° and only 9′ for acute angles. For such
a neat and easily made construction, it is remarkably accurate.

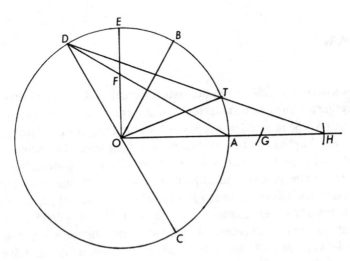

FIGURE 4.93

Untitled Trisection

By S. R. W., 1973

In Figure 4.94, $|OC|$ is one-third $|OB|$, OD is the bisector of angle BOA, and the line through C is parallel to OA. T is found by making $|CD| = |DT|$. This is the same as asserting that

$$\tan\left(\frac{\theta}{3}\right) = \frac{\sin\theta}{2 + \cos\theta}$$

and it is quite a poor approximation. The angle that the author tried to trisect was around 28°, where the error is only about 5′, but it increases rapidly for larger angles.

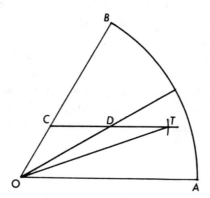

FIGURE 4.94

Any trisection which is fairly accurate for small angles can be improved by first bisecting the angle to be trisected, trisecting that angle, and then doubling it. The error will be reduced by a factor of 4. By bisecting twice before trisecting and then quadrupling the trisected angle, the error would be reduced by another factor of 4, and the construction would be 16 times as good as the original one. And so on: the process could be carried out as many times as you like and the trisection made as accurate as you wish. There is thus no best approximate trisection, since however accurate yours is, I can by bisecting first be four times as accurate as you are. Of course, the physical difficulties in, say, constructing one-thirty-second of a 60° angle, trisecting it, and then doubling it five times would make the theoretical accuracy impossible to attain.

Geometrical Trisection

By A. D. W., 1982

The diagram for this trisection (Figure 4.95) does not show the trisection of a 60° angle, as do almost all of the others, because the construction does not work for a 60° angle, or for any larger angles. This is one of the least accurate trisections I have ever seen. It is constructed by drawing arc CED with center at O, arc FHG with center E and the same radius, and arc IJ with center H and the same radius. T is on arc IJ, $|CD|$ from J. The error is more than 1° for a 33° angle and is in excess of 4° at 48°. The reason for the low quality of the trisection is, I suspect, that for W. the trisection was a mere chip, or splinter, off his system of the universe, to which he probably devoted most of his time. His universe does not have any gravity in it, nor does it have much sense:

> Now, my Solar System was bombarded into space, and strange together all the "bodies" involved in it, it located itself because of the Energy-Motion propulsing it and their respective structural density, each "body" took a place in its own wave of radiation, originating each "body" its own 0-degree orbit, eternally maintaining its 0-degree orbit, and nothing but that 0-degree orbit for as long as its Energy-Motion and its structural density remain consistent.

That is word for word, and it goes on like that for seven single-spaced pages. I had more sense than to write to him, but another mathematician did write:

> The most recent is a [A. D. W.] to whom, in a weak moment, I replied. (He had a proof for a trisection which, while wrong, seemed to show

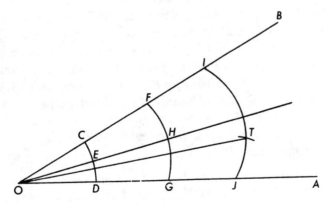

FIGURE 4.95

considerable rationality—or perhaps his Hispanic name lulled me into thinking he might be different.) However, soon it was as usual, mostly now quadratures of the circle. Amazingly he came up with a sequence of "values" for π which were different (his latest value is well below 3). Also, as usual, his work is the "masterpiece of all ages" and came to him "from a greater intelligence."

Mathematicians, do not write to trisectors. Trisectors, do not write to mathematicians.

To Trisect Any Given Angle

By S. W., 1969

Number mysticism:

Line is 10, Time is 9, Square is 8, Area is 7, Divider is 5, Multiplier is 4, Pi is 3, Addition is 2, Point is 1, Circle is 0,

incomprehensibility:

[W.] Law—Time is nine because you can not split a fraction,

and crank's arrogance:

You will have to change all math books.

all appear on this one-page copyrighted sheet, along with a trisection. There is nothing wrong with mysticism: there are more ways than reason of discovering truth, and there are truths which cannot be found rationally (I would tell you about my own mystical experience—in McKeesport, Pennsylvania, in June 1957—except that, by definition, mystical experiences cannot be put into words), but incomprehensibility and arrogance are no aids to trisection.

The construction (Figure 4.96) is long: determine C so that triangle ABC is equilateral and extend CA and CB by one-third of their lengths to D and E. Trisect segment CD at F and join E to F. I is on ED such that $|EI| = |GH|$ and IJ is parallel to CD. T is on the arc GH (with center O) at a distance $|IJ|$ from H. The errors in this construction are so large—39' when the angle is 30°—that I may not have described it accurately. W.'s instructions could have been written more clearly. For example,

Transplit triangle FO^2, J, FO^3 at Y & Z. Compass R, S, transpose at FO^3 & NK, form triangle N, K, FO^3. Increment compass, check to see if line thicknesses have not caused error.

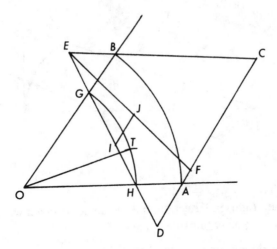

FIGURE 4.96

The angle trisected in W.'s diagram is about 33°, so he may have put the error down to line thickness.

W. also says

> 8 lasers in a web pattern could counter-act gravity, coupled with sound melt stone, etc.

No doubt the failure of his trisection did not get him down: there were plenty of other things to go on to.

Correspondence

Between S. W. W. and N. R., 1983–1984

This file spans five months of time and comprises sixteen items, twelve by W., the trisector, and four by R., a man of great patience and forbearance, and it well illustrates the evolution of a trisection. It is the only such example in this Budget, since I lack R.'s patience and I also know the almost inevitable end of starting to write letters to trisectors. W.'s first communication, headed "The Trisection of Any Angle," contained a diagram in which an angle had been tripled. W. then attempted to go in reverse, but as R. pointed out in his calm and polite reply, there was no way to locate one of the essential points when starting with the large angle. This provoked an annoyed response from W. and, some time later, another trisection. This one was a sure-enough 100

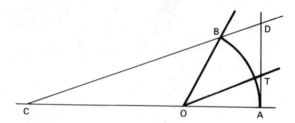

FIGURE 4.97

percent accurate right-on trisection, but the angle trisected was a right angle. R. pointed *that* out, including in his reply

> But please do *not* send me another entirely new way of constructing the diagram and proving that XY is half of YB, because it takes a long time to work through such proofs.

W. took heed. His next try was not an entirely new way: he used the same method as he had used on the right angle on an angle that was *almost* a right angle. This introduced the obvious error of asserting that a line segment was equal in length to the radius of a circle when it was clearly longer, which R., this time with a hint of asperity, brought to W.'s notice.

Then came the breakthrough. A true trisection, of any angle, starting with the angle itself. In Figure 4.97, extend OA backward so that $|OC| = 2|OA|$, erect a perpendicular to OA at A, and find D at the intersection of the perpendicular and CB extended. Trisect DA to find T. Such a simple construction could not be very good, nor is it: it locates the

$$\tan\left(\frac{\theta}{3}\right) = \frac{\sin\theta}{2 + \cos\theta}$$

point found by Father C. and others.

That is the end of the evolution as I know it, since R. gave up and paid no attention to further letters, ignoring ones that started

> How can you ignore the enclosed?

and

> I will try again, but with little hope that you will look at it.

In the meantime, W. had evidently been finding out more about the trisection. One of his early letters contained

Descartes' Proof is now unproofed.

which is odd, since Descartes had nothing to say on the trisection that I know of, but in his last one he was much better informed:

Forget P. L. Wantzel; his proof that it can't be done is flawed.

W. may have gone on to refine his construction or, more likely, to find other people to look at it. His letters sounded as if they had been written by a person irredeemably hooked on the trisection.

The Tri-Secting of Any Angle

By V. W., 1972

A trisection, a duplication of the cube, and a squaring of the circle, all on one sheet, "subscribed and sworn to" before a notary public and copyrighted.

The construction is very simple. In Figure 4.98, $|OC| = |OD| = (1/3)|OB|$ and $|AT| = |CD|$. As might be suspected, it is not very accurate, with errors in excess of $1°$ for acute angles.

W. writes *radius* three times and *radious* three times; *diagonal* is never right, being *diaginal* once and *diagnal* once. Poor spellers make poor trisectors.

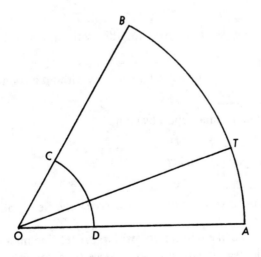

FIGURE 4.98

Untitled Trisection

By H. A. W., 1951

W. went most trisectors one better, giving a construction with straightedge alone, but there were directions like

> The given angle, XOY, is transferred with a carbon sheet to AOD as shown.

to which Euclidean purists would object, since the Greeks had no carbon paper. However, there was also a traditional trisection with straightedge and compass. W. pestered a great many people with his trisection and got a large number of replies. To show you the futility of trying to reason with the typical trisector, here is what W. did *after* he had all of his replies. He summarized them and ended his review with the following:

Ballot for American Mathematicians

Please fill in all blanks; tear off and mail to [H. A. W.], on or before August 1, 1951; or advise him when you will send it. There are four propositions, as follows:

1. Do you agree that the problem is insolvable—in any manner?

Yes _____   No _____

2. Do you agree that the problem is insolvable with compass and ruler, under the classic restrictions?

Yes _____   No _____

3. Do you agree that [W.] has solved it rigidly, with compass and ruler?

Yes _____   No _____

4. Do you agree that [W.] has solved it correctly without a compass?

Yes _____   No _____

May I quote you in my final report by name?

Yes _____   No _____

By the Institution you represent?

Yes _____   No _____

WHY do you think as you do? Please write concisely, clearly and thank you.

It is enough to make one splutter with outrage! The man had numerous replies, all sent by well-meaning people whose time could have been better

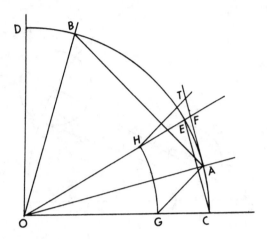

FIGURE **4.99**

spent in almost any other pursuit, and then he had the incredible effrontery to ignore it all and expect American mathematicians to complete his insufferable ballot. Write concisely, indeed! Clearly, indeed! W. is dead and gone to a place where mathematicians are safe from him.

It is a pleasure to report that W.'s trisection is not good at all. In Figure 4.99, BOA is the angle to be trisected and DOC is a right angle, symmetrically placed so that angles DOB and AOC are equal. Angle COE is 30°. AF is drawn parallel to CE, AG is perpendicular to AB, and H is on OE at a distance $|OG|$ from O. HT is parallel to GA. The error is as much as 20′ for angles between 0° and 90°. The construction is equivalent to claiming that

$$\cot(\pi/4 - \theta/6) = 1 + (\sqrt{3} - 1)\tan(\theta/2),$$

which is the same as the construction of A. B. R.

Geometry, Masonry, and Pythagoras

By O. E. W., 1972

This trisection can be located in the 1972 volume of the *California Freemason*. W. found his construction quickly, once he had the idea:

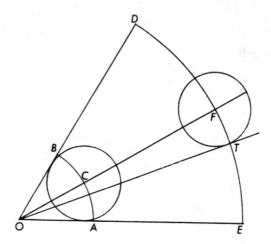

FIGURE 4.100

I idly picked up a book of elementary Trigonometry and the first illustration—a definition of radians—gave me an idea. I took my drawing instruments and in ten minutes I had solved the ancient puzzle.

In Figure 4.100, OC is the bisector of angle BOA, the inner circle has center C and is tangent to the sides of the angle, DE has radius $3|OB|$, and the circle with center F is the same size as the inner circle. T is located as shown. The construction, as could have been expected of one found so rapidly, is not very accurate: the error at 60° is more than 24′ and it increases rapidly. The point located is the same as in the construction of W. D. D., a large coincidence.

I did not know where to find later issues of the *California Freemason* to see whether the trisection was repudiated.

Trisection of Angles

By J. C. W., 1902

There is nothing new under the sun, in particular California eccentricity. Here is the first instance in this Budget of literature being used to trisect:

The problem might have remained unsolved except for a study and analysis of the little poem, "In the Distance," wherein the numbers 3 and 7 seem to coincide in various ways and wherein various other coincidences are demonstrated by the aid of progressive or triangular numbers. Herein was found the key to the solution of the problem.

The poem seems perfectly ordinary: its last stanza is

> Trust not in fame nor wealth to bless;
> Go help the poor and soothe distress;
> Be brave, be true and do your best;
> Do good until with God you rest,
>   In some far wondrous home,
> And all will be as well with thee,
>   Through all the years to come.

But it contains unexpected depths:

The first letter of the alphabet is used as a word and for the commencement of words 33 times: 33 commas are used: there are 33 letters in the longest line and 33 lines preceding it.

The number of letters in the alphabet 26, multiplied by the number of verses 7 = 182, the number of letters in the 7th verse. Seven lines have each 26 letters viz: 2d, 15th, 19th, 24th, 36th, 38th, and 45th: these numbers added together, $2 + 15 + 19 + 24 + 36 + 45 = 179$, the number of letters in the 2d and 6th verses. The sum of all the numbers from 1 to 26 = 351, the number of letters in the 3d and 7th verses combined.

Four pages of the 19-page pamphlet are given over to such things. It is astonishing that W. could have discovered them, and it would be even more astonishing if the author of the poem had put them in on purpose. W. did not say how this numerology led him to his trisection. In Figure 4.101, C is the midpoint of BA, $|AB| = |AD|$, and E is one-third of the way from A to B. F is located so that $|FD| = (1/6)|DA|$ and $GF = 0.7|DF|$. Here perhaps is the mystic influence of 7 and 3. H is constructed so that C, A, and H are vertices of an equilateral triangle and I is the bisector of EH. Draw the line joining I and F; J is on it at a distance $|EG|$ from E. T is at last determined on the arc AB by drawing an arc with center J and radius $|JE|$.

The construction would be very hard to carry out because AD and IF are nearly parallel, and a segment seven-tenths as long as a given segment cannot be constructed quickly. But it is accurate, with no error greater than $2'7''$ for acute angles.

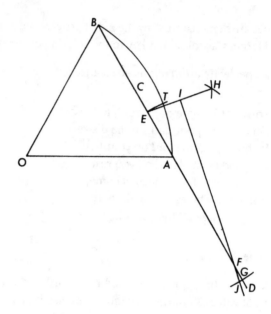

FIGURE 4.101

The [W.] Theorem for Trisecting an Angle Geometrically

By M. A. W., 1945

W. knew that the circumference of a circle is π times its diameter, but

No one knows the exact value of pi. The figures commonly used are 3.1416, but that decimal fraction is not correct. Pi is 3.141592653589793238 plus and plus and plus, on into infinity, so far as man knows. It does not seem wise to teach the youth of today, who will be adults in tomorrow's new world where mechanical instruments can measure to within one ten-thousandth of an inch, that the only way to trisect an angle is to compute with a certain number and then give them the wrong number to start with.

Someone must have told W. that the radian measure of an angle of 60° is $\pi/3$, so that to trisect it, you have only to calculate $\pi/9$. But geometrical constructions are independent of the way that angles are measured. At any rate, the following construction does not depend on π: by bisecting repeatedly (Figure 4.102), divide the arc AB into eight equal parts, thus determining C, D, E, F,

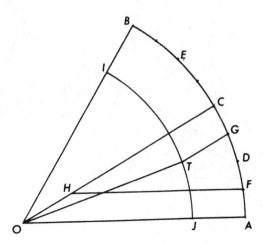

FIGURE 4.102

and G. FH is parallel to OA, and the arc IJ has center O and radius $|HE|$. Drawing a line through G parallel to OC determines T.

For such a clean-looking construction, the accuracy is startling, with no error larger than 1′ 40″ for any angle between 0° and 129°; 60° is trisected to within 38″.

Tri-secting an Angle by Compass and Straight-edge

By E. H. Y., 1931

Y. wrote, in capitals,

NOW FIRST ACCOMPLISHED AND PROVEN BY THE PRO-
CESS OF PURE GEOMETRY, AFTER A FULL THREE THOU-
SAND YEARS (SINCE WHEN THE ANCIENT MAGI KNEW
AND USED IT IN THEIR LODGES), ALTHOUGH LEGIONS
HAVE TRIED FOR IT AND FAILED, AND LIBRARIES HAVE
BEEN WRITTEN ABOUT IT PRO AND CON.

Did the ancient Magi have *lodges*? Did they really sit around in them, trisecting angles? Y. gives no proof that his construction is correct and in fact says

My process is accurate as to any angle up to 90°, but begins to slip beyond that point.

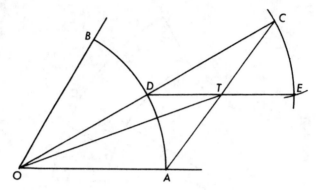

FIGURE 4.103

It is curious that he could recognize the lack of exactness for large angles and yet say that he was precisely right for angles less than 90°.

The construction is yet another version of the

$$\tan\left(\frac{\theta}{3}\right) = \frac{2\sin(\theta/2)}{2\cos(\theta/2) + 1}$$

construction that has been found by others. In Figure 4.103, bisect angle BOA and put C at a distance $2|OA|$ from O. Draw an arc with center D and radius $|OA|$ and locate E by making $|CE| = |DA|$. The trisection point is the intersection of DE and CA.

This ends the Budget of Trisections. Reader, do not add to it.

Index